Praise for D
Green Energy

Solar Electricity Basics is an indispensable
primer for any homeowner or small business
owner considering a photovoltaic system....

— Ann Edminster, author,
Energy Free: Homes for a Small Planet

In *Solar Electricity Basics*, Chiras provides
everything one needs to know about generating electricity
from the sun. I highly recommend this book!

— James Plagmann,
AIA + LEED-AP, Green Architect

Wind Power Basics is a wonderful source of basic
wind info and a must-have for small-wind newbies!
As a trainer of small-wind installers, I will
recommend this book to all my students.

— Roy Butler, NABCEP Certified PV Installer®

With *Wind Power Basics*, Chiras shares what he knows
in a very accessible way. Get this book onto your kitchen
table and spend some pleasant time with it!

— David Wann, coauthor, *Affluenza* and
author, *Simple Prosperity*

GREEN
TRANSPORTATION
Basics

DAN CHIRAS

with Dominic Crea

Illustrations by Anil Rao, Ph.D.

NEW SOCIETY PUBLISHERS

Cataloging in Publication Data:
A catalog record for this publication is available from
the National Library of Canada.

Cover design by Diane McIntosh.
Cover images: Car image courtesy of Zap! Electric Vehicles: Model: Alias,
Zapworld.com. Other cover images © iStock, alphacat (background leaf)
Diane Labombarde (electrical cord).

Printed in Canada. First printing July 2010.

Paperback ISBN: 978-0-86571-619-3
eISBN: 978-1-55092-460-2

Inquiries regarding requests to reprint all or part of *Green Transportation* Basics
should be addressed to New Society Publishers at the address below.

To order directly from the publishers, please call toll-free (North America)
1-800-567-6772, or order online at www.newsociety.com

Any other inquiries can be directed by mail to:

New Society Publishers
P.O. Box 189, Gabriola Island, BC V0R 1X0, Canada
(250) 247-9737

New Society Publishers' mission is to publish books that contribute in fundamental ways to building an ecologically sustainable and just society, and to do so
with the least possible impact on the environment, in a manner that models this
vision. We are committed to doing this not just through education, but through
action. Our printed, bound books are printed on Forest Stewardship Council-
certified acid-free paper that is **100% post-consumer recycled** (100% old growth
forest-free), processed chlorine free, and printed with vegetable-based, low-VOC
inks, with covers produced using FSC-certified stock. New Society also works to
reduce its carbon footprint, and purchases carbon offsets based on an annual audit
to ensure a carbon neutral footprint. For further information, or to browse our
full list of books and purchase securely, visit our website at: www.newsociety.com

Library and Archives Canada Cataloguing in Publication

Chiras, Daniel D.

Green transportation basics / Dan Chiras ; illustrations by Anil Rao.

Includes index.
ISBN 978-0-86571-619-3

1. Alternative fuel vehicles. 2. Electric vehicles. 3. Automobiles-- Technological
innovations. I. Title.

TL216.5.C45 2010   629.22'9   C2010-903411-2

NEW SOCIETY PUBLISHERS

Mixed Sources
Product group from well-managed forests,
controlled sources and recycled wood or fiber
www.fsc.org Cert no. SW-COC-000952
© 1996 Forest Stewardship Council

Contents

Books for Wiser Living
recommended by *Mother Earth News*

Today, more than ever before, our society is seeking ways to live more conscientiously. To help bring you the very best inspiration and information about greener, more sustainable lifestyles, *Mother Earth News* is recommending select New Society Publishers' books to its readers. For more than 30 years, *Mother Earth* has been North America's "Original Guide to Living Wisely," creating books and magazines for people with a passion for self-reliance and a desire to live in harmony with nature. Across the countryside and in our cities, New Society Publishers and *Mother Earth* are leading the way to a wiser, more sustainable world.

CHAPTER 1

GREENING TRANSPORTATION

U nless you've been in a coma for the last decade, you very likely won't need convincing that the world needs to find a way to "green" its transportation system — that is, to make it more environmentally sustainable. Much more sustainable. Global climate change, our heavy dependence on declining oil reserves, and high fuel prices should be enough to convince even the most stalwart opponents of all things green that we must do something — and soon — to create a leaner, greener mode of transportation.

I won't recite the catalogue of facts and figures here to make the case to you. You're reading this book because you very likely already know them and feel the need to make changes. Chances are you are responding to a deep conviction that it's time you did something — or perhaps more — to green your own transportation. This book will help enormously. It covers three basic areas: (1) ways you can drive less "fuelishly," (2) green vehicles options like plug-in hybrids, and (3) green fuels such as biomethane, ethanol, and hydrogen.

While my emphasis since the early 1970s, when I became actively involved in energy efficiency and renewable energy, has been on residential energy efficiency and renewable energy, I have also studied green transportation extensively. It's been one of several "side passions" of mine since 1971. It all started when I was nearly 20. I had just gotten married. After the ceremony, my wife and I drove from western New York State where I grew up, back to college in eastern Kansas, where we were both finishing our senior year. My parents had given us a Chrysler 300 as a wedding present. It was a large, sprawling ocean liner of a car! As we drove west along Interstate 70, we watched in horror as the needle on the gas gauge plummeted. We could actually see the needle move as we were driving. A few days after we returned, we traded in the monster for a fuel-efficient Volkswagen Beetle.

My next vehicle, which I bought after graduating with a Ph.D. in reproductive physiology, was a relatively efficient Datsun pickup truck. Efficient as it was, I chose to commute by bus or bicycle to the University of Colorado in Denver for my first teaching job. Every car I've driven since then has been fuel-miserly. Today, I drive a fuel-pinching Toyota Prius and, more recently, an electric Chevy S-10 pickup truck. My brother-in-law and I converted the former gas-engine truck to electricity at The Evergreen Institute's Center for Renewable Energy and Green Building, my educational center in east-central Missouri (Figure 1.1).

In addition to pursuing ways to green my own transportation, I've attended workshops and numerous lectures on green cars and green fuels over the years and read every article that crossed my desk as well. This book is the culmination of much of my informal research — my green car hobby. Its focus is on *personal* transportation, not transportation *systems*, so it doesn't include much information on trains and buses or other very green options. Although I briefly discuss mass transit and car-sharing programs, I focus primarily on what *you* can do to green up your act.

Fig. 1.1: *My latest, most sustainable transportation: a former gas-engine truck that my brother-in-law and I converted to electricity. This vehicle is used for short trips to town and around the farm and The Evergreen Institute.*

Before I start, though, I will discuss some criteria to consider when attempting to green personal transportation. These rules of the road, so to speak, will help you discern what makes a truly sustainable fuel or vehicle. With these guidelines, you can proceed quickly, not wasting time and energy on ideas that really have no long-term future.

Guidelines for Sustainable Fuel Transportation

When it comes to green transportation, you'll find that there's a plethora of options. When you're choosing among them, it is important to make selections that are socially, economically, and environmentally sustainable. Green fuels and vehicles, that is, must make sense from all three perspectives to be truly sustainable. That's the first rule. If an option doesn't make sense from these perspectives, it is not worth our time or effort. Period.

With that principle in mind, we will begin with criteria by which we can judge green fuels, such as hydrogen and ethanol.

First, when considering green fuels, the most sustainable are those that have a positive net energy yield. *Net energy* refers to the energy we obtain from an energy resource after subtracting the energy it takes to make it — that is, to extract, refine, and process a fuel. The higher the yield, the better. The most sustainable fuels are therefore those with the greatest net energy yield.

Second, a sustainable fuel must be clean — in every possible way. That is, it must be produced in ways that do no harm to people or the environment — for example, to be sustainable, biofuels must come from crops that are sustainably grown and harvested. When burned or consumed, a truly green fuel should produce little, if any, harmful pollution. The cleaner and more environmentally benign, the better. We can't build a sustainable future by turning to fuels whose production and consumption poison people and other living creatures that grace our world or to fuels whose production and use threaten our atmosphere, waters, climate, and ecosystems, which are the life support system of the planet.

Third, for a fuel to be sustainable it must be abundant and renewable. We can't build a long-term transportation system on short-term fuels. That's the bind we're in now. Gasoline- and diesel-fueled vehicles rely on a resource (oil) that's quickly going the way of the dinosaurs, and is, in the process, causing severe economic turmoil. For economic stability, we need fuel we can count on forever. Building any transportation system based on fuel that's going to give out in the near future is a waste of very precious energy, resources, and time.

Fourth, we must select options that are affordable — if not now, then clearly in the near future as economies of scale kick in or as improvements in production processes drive costs down. If subsidies are required now, fine, but in the long run, affordable fuel is vital to the long-term economic health of nations and their citizens.

The rules of the road for green *vehicles* overlap nicely with my four *fuel* guidelines. First and foremost, green vehicles must be powered by sustainable fuels. They should also be durable and safe. We don't want people dying on the highways in their quest for a greener world.

Truly green vehicles must be made from recyclable materials and designed for ease of recycling. Moreover, we must establish the infrastructure to ensure that they *can* and *will* be recycled when their useful lives are over. Even better, green cars and trucks should be made from renewable resources, for example high-strength plastics made from chemicals derived from sustainably harvested plants.

In crowded urban centers, mass transit will very likely be the most sustainable form of transportation. Such systems service large numbers of people with much less impact on the environment than single occupancy vehicles. The amount of fuel consumed per passenger mile is much lower, as is the amount of pollution produced. Still, people will very likely cling tenaciously to their cars and trucks — even if they choose to rent a car from a local car cooperative or a private company that provides the convenience of private transportation without the hassles of ownership. Rest assured, private passenger vehicles will not disappear quickly from the urban landscape in many countries, making green fuels and green cars imperative.

Keep Your Brain Engaged

Over the years, I've heard many cockeyed schemes aimed at greening transportation. While many new ideas appear great at first blush, most don't hold up to scrutiny. As you no doubt have found out, it is very easy to become swayed by novel fuel sources or novel types of vehicles, like air-powered cars that the ill-informed media proclaim will revolutionize transportation. Enjoy the creative ideas, but employ your critical faculties. The most important question you can ask when presented with a new fuel or vehicle is this: is it truly sustainable? Sustainability should be measured by the criteria just presented, among them net energy efficiency. Inquire about the net energy efficiency of exciting new developments. Does it take more energy to make a fuel or vehicle than you get out of it? Don't be swayed by pro-oil interests who muddy the waters with falsehoods (Figure 1.2). Critics of green fuels are fond of pointing out the seemingly unfair subsidies required to support production of green fuels and new green vehicles. What they're less forthcoming

Fig. 1.2: *Don't be fooled by lies. Contrary to the media's representation, ethanol has a much higher net energy efficiency than gasoline. When all the energy surrounding production and delivery is taken into account, we get a lot more energy from ethanol than we do from gasoline.*

DAN CHIRAS

about, however, is that gasoline and diesel require huge amounts of energy to extract, refine, and transport to market. The net energy efficiency of gasoline is very likely under *one* — meaning you get less energy out of a gallon of gasoline than it takes to make it. You won't find oil companies passing that information around. Moreover, the oil industry and their legions of followers don't say a whole lot about the amazing subsidies that continue to flow into company coffers. So don't let critics win the debate with their lies!

Remember that we're in an experimental phase on the path to a green energy future. Lots of new ideas are emerging. Some will work; others will fail miserably. Some will work because there are powerful lobby groups that stand to benefit from them. Seemingly good ideas will fail because they lack financial or political support. It will take time to sort through the options, to see what works and what doesn't. In the meantime, enjoy the ride.

To speed up the evolutionary process, though, we should apply some sensible guidelines to help us judge what is truly green. After all, we were endowed with marvelous brains, so let's use them. On a personal level, while you may not be able to influence public policy, you can make wise choices for yourself based on your understanding of what makes a fuel and a vehicle truly sustainable.

With these ideas in mind, let's get started where all journeys of sustainable transportation should begin, with what you can do *right now* to green your wheels. That's the subject of Chapter 2.

CHAPTER 2

EASY GREEN:

CHANGING DRIVING HABITS, AND OTHER SIMPLE MEASURES TO GREEN YOUR MACHINE

When most people ponder ways to green their personal transportation, they contemplate grand schemes like futuristic hydrogen-powered cars. Or they may contemplate purchasing a hybrid or electric car or truck or converting their diesel vehicle to run on straight vegetable oil. While most of these are legitimate options, they require considerable thought and, in some cases, a substantial monetary investment.

If these and other green transportation options are more than you want to pursue right now, don't let your dream of greening your transportation idle. There are a great many ways you can reduce fuel consumption and help build a sustainable transportation system — right now, at little or no cost. You can, for instance, increase your use of mass transit or join a carpool or vanpool to commute to work. You can ride a bike or walk when going to work or when running errands near your home or office. If you live in a major city, you could sell your car and subscribe to a service like Zipcar that rents cars to members when they need one, saving members a ton of money (Figure 2.1).

You could even embark on the nearly unspeakable act of altering your driving habits. It may be hard to believe, but driving

Fig. 2.1: *Cars like this one supplied by Zipcar can be rented by city dwellers on an as-needed basis, greatly reducing the cost of personal transportation and the environmental impact of private vehicle ownership.*

your existing vehicle more efficiently can achieve the same level of fuel savings you could achieve by pursuing more costly options, like buying a more fuel-efficient car or truck. Moreover, there are a number of simple things you can do to improve the fuel efficiency of your vehicle, like keeping the tires properly inflated. Driving more efficiently and improving the efficiency of your car save gas and money, reduce pollution, and reduce the wear and tear on your vehicle. And, on top of that, fuel-efficient driving can make you a *safer* driver.

In this chapter, you will find a number of simple, practical suggestions that can help you improve fuel economy.

Wise Driving Habits

A couple of years ago, I took a road trip with some colleagues from Denver, Colorado, to Greensburg, Kansas, with the hopes of building some energy-efficient homes. The homes were to serve as a model to the community and the rest of the world in our collective effort to combat global warming — which had no doubt contributed to the devastating 1.7-mile-wide tornado that ripped through Greensburg in 2007, wiping out 95% of the homes and killing 11 people. For most of the trip, I was behind the wheel, with my colleagues chiding me for driving 60 miles per hour (mph) on the wide-open highways. After a while, I tired of their joking and said, "Hey guys, I don't get it. You're complaining because we're driving at a slower speed, which increases mileage. That, of course, reduces carbon dioxide emissions, which reduces global warming. Global warming, of course, contributes to violent storms like the one that destroyed Greensburg."

There wasn't a word of rebuttal.

One of my friends, a green architect I work with frequently, later said he was going to drive more slowly in the future — and has kept his promise, too!

Over the years, many environmentalists — the people who presumably care the most about global warming and other environmental issues — have shamelessly admitted to having a "lead foot." One after another has confessed to me that when it comes to the need for speed, they're like most Americans: when they hit the road, getting from point A to B in the least amount of time is their number one priority. If environmentalists can't slow down to save the Earth, how can the rest of us?

If you're one of those people for whom speed is a given, you should know what your aggressive driving habits are costing you. Aggressive driving dramatically lowers fuel economy — both in the city and on the highway. In the city, it easily reduces fuel consumption by 5%. On the highway, it can lower fuel economy by up to 30%. If you're getting 300 miles per tank of gas, driving more sensibly could add another 75 miles to your range. That means you can go longer between fill-ups, saving time and money. If you're spending $2,000 a year on gasoline, an easily achieved 25% savings is a whopping $500. If you earn $25 per hour, you'll be able to work 20 hours a year less! In addition, more sensible habits reduce the release of pollutants and our dependence on foreign oil. So, how can you tame those aggressive habits?

Maintain a Constant Speed. When driving on the highway, use the cruise control to maintain a constant speed. Set it at the speed limit, or a little below, then sit back and relax. Relying on cruise control is a simple measure that saves gas (and reduces strain on the muscles of your leg and foot). According to the experts at Edmunds.com, cruise control can improve fuel mileage by 7–14%, except in mountainous terrain.

If you prefer to control speed yourself, be sure to keep a steady foot on the gas pedal. This method is more challenging and results in more muscle strain, especially on long trips. And few people find it easy to maintain a constant speed.

While maintaining a constant speed is ideal, it's not always practical. There are times when you need to speed up. If you must accelerate, it helps to wait until you are on a downhill section. Let gravity assist your car's engine — and help save gas. You'll be amazed at how many declines there are on seemingly flat highways, even in the Midwest.

Avoid Tailgating. When driving in traffic, be sure to leave sufficient space between your car and the car in front of you. If the driver in front of you speeds up, don't feel compelled to follow suit. Chances are the driver will soon slow down. If you are following too closely, you'll need to step on the brake too, wasting all that fuel you just spent speeding up. In essence, you've just allowed the anonymous driver in front of you to be in charge of your own gas mileage.

Avoid Jackrabbit Starts. Many people accelerate from a stop light or stop sign as if they're on the starting line of the Indy 500. With the pedal pressed to the floor, they dart off like a race car driver intent on winning the trophy. These are called jackrabbit starts.

Unfortunately, quick starts waste a tremendous amount of fuel, especially when traveling on roads with frequent stop lights or stop signs. When the light turns green, resist that urge to step on the gas.

You don't have to be the first off the line. Chances are, the few seconds you think you are saving will be lost at the next stop light anyway. The more conservative driver in your rear view mirror will very likely catch up to you, reaching the same destination having used a lot less fuel. By driving more efficiently, you'll spend less time at the gas pump during your life. Any time you "save" by driving like a maniac will eventually be offset as you mindlessly wait at the gas pump for the tank to fill. And you'll spend more of your life working to pay for the fuel you burn unnecessarily.

Slow Down, You Move too Fast. Simon and Garfunkel's advice is as relevant to driving as it is to us in our everyday lives. Numerous studies show that fuel mileage decreases dramatically the faster a car goes. In most cases, fuel mileage begins to plummet at speeds over 50–55 mph. One study by *Consumer Reports*, for instance, showed that fuel mileage fell 12.5% when a car's speed increased from 55 to 65 mph. It dropped another 12.5% when speed increased from 65 to 75 mph. Why does speed increase fuel consumption?

The main reason is something engineers call *aerodynamic drag*. This is the force on a car that resists its motion through air. One of the sources of aerodynamic drag is skin friction. Skin friction occurs between the air molecules and the exterior of a car. Because the skin friction is an interaction between a solid and a gas, its magnitude depends on the properties of both. The smoother the surface, for instance, the lower the drag. The denser the air, the greater the drag.

Drag is also generated by the resistance air molecules pose to an object in motion through it. This source of drag depends on the shape of the object (it's called *form* drag). In cars, then, drag is affected by the frontal area of a car, its design (how sleek and aerodynamic it is or isn't), and the density of the air.

As a general rule, drag increases as a function of the square of the speed. That means if the speed doubles, drag quadruples. And, while drag is important, we're even more interested in power requirements and fuel consumption. As a general rule, the horsepower required to move a vehicle forward increases as the *cube* of the speed. If the speed doubles, the horsepower required to move the vehicle increases five to six times. For example, a hypothetical medium-sized SUV that requires 20 horsepower at 50 mph might require 100 horsepower at 100 mph. Does that mean the fuel consumption increases accordingly?

No, things are a bit more complex than that. While power increases as a function of the cube of the speed, fuel consumption doesn't necessarily increase in step. That's because engines are very complicated machines and efficiency is a function of other factors besides drag. For one thing, most engines normally operate most efficiently near 80% of their full throttle.

So what's the bottom line? Driving at high speeds dramatically increases fuel consumption. Driving more slowly substantially reduces fuel consumption. Those who drive more slowly consistently get better gas mileage than the ratings provided by the EPA on every new vehicle. You can take that to the bank.

How much fuel economy declines depends on the design of the vehicle and the engine, but you can count on a substantial decrease at speeds over 60 mph — from 7–23%, depending on the vehicle. The faster you go, the lower the fuel economy. If your car averages 30 miles per gallon (mpg) at 60 mph, fuel mileage could plummet to 23 mpg — or even lower — at 75 mph. You're essentially throwing away about one fourth of the fuel that you put into the tank, fuel for which you paid dearly.

Because engines vary with respect to efficiency, and the aerodynamics of vehicles also vary considerably, the sweet spot for each car varies. For example, Jeep Cherokees operate most efficiently at 55 mph, while Toyota 4Runners operate most efficiently at a slightly slower 50 mph. Some stick shifts with small engines operate most efficiently at around 35 mph!

It may take some experimentation to figure out your vehicle's sweet spot. If your car comes with a fuel economy gauge (like the Ford Fusion or Chevy Cobalt), it's relatively easy to find the sweet spot. To do so, select a very flat section of highway (a 2–5 mile section is best) and let 'er rip. Try 50 mph for 5 minutes, then turn around and go back to the starting point. Record the fuel economy. Now drive the same section at 55 mph. Record your fuel economy. Repeat at 60, 65, 70, and perhaps even 75 mph, provided high speeds are permitted, of course.

Some cars, like the Toyota Prius and Honda Civic Hybrid, have fuel mileage gauges that indicate the immediate mileage but also keep track of long-term fuel mileage, which will make your experiment run a little smoother. Prius owners have found that this car performs best at speeds below 50 mph, perhaps even as low as 40 mph. I notice a substantial increase in mileage in my 2004 Prius when cruising at 55–60 mph compared to 65 or 70.

I'm *not* advocating dangerously slow speeds, especially on major highways, but you should know the facts. For most cars, you'll find that fuel mileage decreases fairly substantially at speeds over 60 mph.

Shift Properly. Most passenger cars and light trucks these days are automatic that is, they shift on their own. But there are a number of vehicles on the road that require the driver to shift gears. As a rule, vehicles with manual transmissions are more fuel efficient than those with automatic transmissions. Those who drive cars with manual transmissions, however, can improve their fuel economy even more by paying attention to when they shift and the gear they are operating in. Shifting a few hundred rpm lower than you normally do, for instance, can save on fuel. Operating in fourth gear when the car should be in fifth gear is wasteful, too, as it increases the rpm and uses more fuel. So pay close attention. If the engine's racing, you should be in a higher gear.

Selecting the right gear depends on the engine, the transmission, and the driving conditions. If you're going up a hill, for instance, you may want to shift into a lower gear to provide more power. On the open highway, though, your car will generally operate most efficiently in the highest gear.

When stopped, watch your foot on the gas pedal. There's no need to race the engine. And when taking off, press on the gas pedal slowly. Give the car only enough gas to maintain the idle rpm as you let off the clutch. As any experienced manual shift driver knows, there's no need for a heavy foot when just starting off.

Smart Braking. As you approach a stop sign, a stop light, or highway congestion, you can save gas by removing your foot from the accelerator pedal well before you need to stop. This

Shopping Tip

When buying a new car, look for one with a *continuous variable transmission* (CVT). It does not shift from one gear to another, but changes the gear ratio constantly so you're always in the right gear.

allows your vehicle to decelerate gradually, essentially gliding to a stop. Planning ahead for braking in stop-and-go traffic saves a lot of fuel over the long haul. There's no sense speeding up to a stop light, then slamming on the brakes. Although it's a common habit, it's extremely wasteful. As an additional suggestion, if you drive a manual transmission, you may want to slip the car out of gear as you glide to a halt or downshift to use your engine to brake the car.

• **Get Up and Go.** One time when impatience pays is in the morning when you're on your way to work or school or off to run errands. Many people are under the false impression that letting their car idle for several minutes helps the engine. Although gas engines should be given a short period to warm up, 30 seconds is usually sufficient. Letting the car idle for 5 or 10 minutes while you pack your briefcase or slam down your coffee wastes a lot of fuel.

The best way to warm up a car in the morning is to start your engine, let it run for 30 seconds, then take off, driving slowly until the engine reaches the proper operating temperature. It only takes a few minutes. If your car has an engine temperature gauge, you can tell by watching it.

Shut 'er Down. You can also reduce fuel consumption by reducing engine idling at stop lights, railroad crossings, and fast-food drive-up windows. If you expect your car will idle longer than 10 seconds, save gas by turning the engine off and then restarting when traffic starts up. Don't go crazy on this, as stopping and starting an engine too frequently may result in excessive wear and tear on the starter and the car's wiring.

Smart Trip Planning Improves Fuel Efficiency

How you drive affects fuel consumption, but good trip planning can also help, often substantially. To reduce fuel consumption, consider better planning.

Combine Trips. To save gas, I routinely maintain a list of errands I need to run each week, then I run as many as I can in one trip. Doing so decreases the number of miles I drive each week and thus reduces the amount of fuel my car burns. Combining trips, in turn, reduces emissions and pollution and wear and tear on my car, saving additional money. And it saves a lot of time.

Combining trips not only cuts vehicles miles traveled (VMT), it also boosts the fuel economy of a car by reducing the number of engine starts and the amount of time a car spends "running cold." As you may know, starting a car uses a lot of fuel, so reducing engine starts reduces waste. Cold engines also operate less efficiently than warm engines, so combining trips, especially in cold weather, increases the fuel efficiency of your vehicle. Because warm engines operate more efficiently, they produce less pollution, too.

My advice is to combine errands into one or two trips a week, whenever possible, especially if each of the short trips involves cold starts. Be especially diligent about combining trips in cold weather. And, whenever possible, eliminate short trips followed by long intervals during which the engine cools. For nearby errands, you may want to consider walking, riding a bike, or taking a bus.

To avoid frequent errands, consider buying staples and other supplies, such as toilet paper, in bulk. Having a backup supply will cut down on "emergency runs." A pantry full of food is a great energy-saving resource.

Plan Your Route Carefully. You can reduce fuel consumption by planning routes that require the least amount of time on the road and the fewest number of stops. As you no doubt have found, you spend a lot of time waiting for traffic to clear to make left turns. When I run errands, I try to plot paths that require only right turns. (It's really quite simple.) This strategy is used by United Parcel Service (UPS). When their drivers leave in the morning, they're assigned deliveries in an order that reduces the number of left turns. This simple strategy saves UPS a fortune in fuel costs and driver time. You'll be surprised at how much time and money it can save you, too.

When plotting routes, also remember that it's usually best to drive on the highway rather than on city streets. Highways generally have less stop-and-go traffic, which reduces idle time and fuel waste. (One exception may be during "rush hour" traffic, perhaps the greatest misnomer in the human language, as traffic often moves at a snail's pace and frequently comes to a complete halt.)

Finally, when shopping, avoid the impulse to secure the closest parking spot. Shoppers waste a lot of gas cruising parking lots of malls searching for prime parking spaces. To save time and fuel, park on the periphery. The exercise will be good for you, and you'll save a lot of fuel — and probably some time — in the long run.

Modifying Your Vehicle to Improve Fuel Efficiency

Smart driving and trip planning can save a lot of costly fuel and help reduce pollution, but there's much more you can do to improve fuel efficiency.

Remove Roof Racks and Roof Carriers. One way to make your vehicle more fuel efficient is to remove unused rooftop racks. Racks for skis and boats and rooftop luggage carriers are handy, but they decrease fuel economy by increasing aerodynamic drag, which slows vehicles down and reduces gas mileage.

Rooftop racks and carriers counteract the millions of dollars auto manufacturers have spent in recent years to make their vehicles more aerodynamic. If possible, remove racks and rooftop carriers when they're not being used. Doing so could increase your fuel mileage by 1–5%. Even flags, banners, and toys attached to antennas can increase drag and reduce fuel mileage. Remove the banner of your favorite football team after the Sunday game and save gas!

Air Conditioner or Open Windows? Driving with the windows open increases drag and lowers fuel mileage, so many people drive with their air conditioner running. Unfortunately, operating a car's air conditioner increases the work an engine must perform, which increases fuel consumption. So what's a smart driver to do?

The rule of thumb is to drive with the windows open at lower speeds (below 50 mph) unless it is sweltering hot, of course. At higher speeds (over 50 mph), close the windows and turn on the air conditioner to stay cool. Studies show that when traveling at highway speeds, fuel efficiency increases (but only slightly) if the windows are rolled up and the air conditioner is running.

Repair Body Damage. The aerodynamics of cars are compromised by dents in the body, so repair body damage as soon as possible. Keep your car clean, too, as mud on the body can increase drag and reduce fuel mileage.

Slow Down on Wet Roads. Road conditions also affect fuel economy. Water on the highway, for instance, increases the resistance the tires encounter, reducing fuel economy. Although you may not be able to avoid driving when roads are wet, consider slowing down. This is not only safer (it reduces

Advice on Air Conditioning

Whenever possible, avoid using the air conditioner in stop-and-go city traffic. Operating the air conditioner causes the engine of a car to work harder, causing it to consume a lot more fuel.

hydroplaning), it also reduces resistance to forward movement, thus saving gas.

Lose the Weight. Many people routinely carry a lot of additional weight in their cars — in rooftop carriers or in the back seat or trunk. Because the weight of a vehicle affects how hard an engine must work to accelerate or maintain speed, extra weight reduces fuel economy. An additional 100 pounds, for instance, can decrease fuel consumption by 1–2%.

Whenever possible, remove unnecessary cargo from your car or truck. I thoroughly unpack my car after each trip, putting items away that I don't need to haul around on errands or short trips. Be sure to remove your chains and ice scrapers at the end of the winter.

Some energy conservation zealots have been known to remove unused seats in their vehicles (especially vans) to reduce weight. Although this may be too extreme for most people, you can easily trim the weight of your car by emptying your trunk or back seat or the bed of your pickup of unnecessary items. And if you've been looking for an excuse to lose those additional pounds, this might be just what the doctor ordered!

Keep Your Car Tuned. A well-tuned car is an efficient one. Keeping your car tuned can increase fuel mileage by up to 4%. A properly tuned engine also maximizes power. (Tell your mechanic you do *not* want efficiency measures disabled when your car is being tuned for power.)

Car Buying Tip

While excess cargo weight is important, overall vehicle weight is an even more important factor in determining fuel economy. When buying a car, pay attention to vehicle *curb weight.* Purchase the lightest, safest car you can.

When a car is tuned, mechanics typically replace the fuel and air filters to improve the efficiency of the engine. (A very dirty air filter can reduce fuel mileage by 10% and can make an engine stall when it's idling.) Be sure to replace the fuel and air filters according to the manufacturers' recommendations.

Be sure your mechanic checks the oxygen sensors, engine emissions system, and evaporative emissions control systems when tuning your fuel-injected vehicle. (The vast majority of the cars and light trucks on the road today are fuel injected.) Resist the temptation to ignore warning lights. Take your car in for service if the "check engine" light comes on. This is usually an indication that there is a problem with one of the vehicle's sensors. All are important to the proper and efficient functioning of the computer-controlled engines of modern vehicles. A

Tire Buying Tips

When buying new tires, be sure to ask the supplier for their most fuel-efficient tires. They are known as "low-rolling resistance tires." These tires are slightly harder than others, which reduces friction between the tires and the surface of the road. Fuel-efficient tires often come with a new vehicle — they're the same tires the manufacturer puts on its cars to rate the car model for fuel efficiency, and manufacturers use every trick they can to boost their fuel economy ratings. If you are satisfied with how they wear, consider buying the same model when it comes time to replace them. I've been disappointed with the performance of the original tires on one of my cars in the past. Although they did help reduce fuel consumption, they didn't last long. In fact, I didn't get a full season out of them. (I can't say for sure, but I suspect the tread was too thin.) Whatever you do, be sure new tires ☛

damaged oxygen sensor, for instance, may throw off the fuel mixture, decreasing fuel mileage by 20% or more.

When your mechanic tunes your car, ask him or her to look for fouled fuel injectors. Clean or replace them when necessary.

Maintain Proper Tire Pressure. You've heard the advice a million times: maintain proper tire pressure. According to several sources, properly inflated tires can improve the fuel mileage of a car or truck by up to 3%. Although this is a small savings, when you add it to other measures like saner driving habits, combining trips, and keeping your car tuned, the cumulative effect can be quite substantial.

Your tires can lose up to one pound per square inch (psi) in pressure per month. Air pressure inside a tire also decreases

will last. Ask the supplier for projected wear data. Michelin and Goodyear have recently released fuel-efficient tires that could save a significant amount of fuel over the lifetime of a set of tires. I put the Goodyear tires on my Prius and noticed an immediate increase in fuel economy.

When buying tires, select the narrowest possible tires that will work with your driving style (aggressive vs. mild-mannered, for instance). Aggressive driving requires wider tires. Narrow tires reduce the frontal area, the amount of surface area that contacts the wind. Reducing the frontal area reduces aerodynamic drag. Narrow tires are therefore typically more fuel efficient. Avoid the tendency to oversize tires. If a tire supplier is out of the right size, they may try to sell you a wider tire. Remember, however, that narrow tires provide less traction, so you don't want to go overboard. ∎

when outside temperatures plummet. Cold temperatures cause the air inside the tires to contract and pressure to drop, reducing the efficiency of the car.

Most experts on fuel efficiency recommend that car owners check the air pressure in their tires at least once a month, preferably weekly. When you get a new set of tires installed, ask what the recommended tire pressure is. Write the number down. (Your owner's manual should also list proper tire pressure.) You may want to consider buying a portable electric compressor to fill your tires. It's a lot easier than running to the gas station once a month to top off your tires.

Besides saving fuel, proper inflation also helps avoid uneven wear and tear on the tread of a tire, saving you money in the long run.

Change Your Oil on Schedule. To improve fuel economy, change the oil of your car frequently. Clean oil allows the engine to operate more smoothly.

You may receive conflicting advice about when to change your oil. The 3,000-mile oil change recommended by the guys who are profiting is excessive according to several car experts; 6,000 to 8,000 miles is probably a good range. (Extending oil change intervals too much can damage your engine and could reduce fuel mileage.)

Changing oil on a set (but reasonable) schedule can improve fuel mileage by an additional 1–2%. That's because oil gets dirty and its chemical composition changes with use. Both dirt and chemical changes in the oil increase its viscosity (thickness). Increased viscosity means the engine operates under a heavier load — so it's harder for the internal components like the pistons to move. This, in turn, reduces fuel efficiency.

While you are at it, you may want to consider adding an oil additive to your oil, which is available at auto supply stores, or using a low-friction synthetic oil. Both can improve fuel mileage.

Because synthetic oils have a lower viscosity than conventional engine oil, they can increase fuel economy, on average, by 5% (in some cases by up to 15%). Also, because they are less subject to chemical breakdown, they will outlast conventional oil and won't need to be replaced as often.

Unfortunately, synthetic oils are substantially more expensive than standard engine oil — three to four times more! The higher cost, though, may be offset by less frequent oil changes — you can easily go 8,000 miles between oil changes. Some synthetic oils don't need to be changed for 25,000 miles.

Synthetic oil is better on a number of other levels as well. It keeps the engine cleaner through improved sludge and varnish protection and reduces engine wear at high temperatures. It also protects engines that are running under severe conditions at high temperatures. Synthetic oil provides better cold-temperature starts with faster oil flow at ignition, and thus improves fuel efficiency.

Fill 'er up. Some people operate their vehicles near empty much of the time, putting only a few dollars worth of gas in at any time. Unfortunately, running on fumes can reduce fuel economy. Why?

Gas splashes around in the fuel tank as we lumber along highways and byways and can splash away from the opening through which gasoline flows to the fuel line. If your tank runs low, splashing means that the engine does not receive a steady supply of gasoline. This, in turn, can reduce the efficiency of your engine. Whenever possible, try to keep your tank at least one third full at all times.

Keep a Log. I calculate fuel economy in my head every time I fill up my car, and have done so for years. Unfortunately, my mind is not a perfect calculator. I round up or down or drop numerals to the right of the decimal point to make the math easier.

If you want to monitor your fuel mileage to see if changes you've made are working, consider starting a log. Keep track of the number of gallons you add each time you fill your tank, write down the odometer reading, and calculate the number of miles you traveled. Keep track of fractions (fuel) and decimals (mileage). If you have the time, you may even want to put this information on a spreadsheet. Or if you have an iPhone, yes, "there's an app for that." Actually, several of them. One is called Vehicle Log Book.

Without a precise way of tracking mileage and fuel consumption, you'll never know for certain whether you are reducing fuel consumption or increasing it.

Conclusion

Clearly, there are a lot of easy, inexpensive ways to improve fuel economy in cars, trucks, SUVs, and vans. If you are serious about doing your part to reduce our dependence on foreign oil and combat climate change and other environmental problems, rest assured you don't have to lay down $30,000 for a new hybrid. You can start today by becoming a more fuel-conscious driver.

Why not join the tiny band of responsible drivers who are cruising the highways, driving at or below the speed limit on their carefully planned routes, combining errands whenever possible, and keeping their cars tuned, saving the world one gas tank at a time?

CHAPTER 3

THE HYBRID REVOLUTION

Hybrid cars and trucks are all the rage right now. Even locomotive manufacturers have produced hybrid models that dramatically improve fuel economy. Although not all hybrids are created equal, this technology could help us green our transportation system. It may be an excellent choice for many readers.

In this chapter, we'll explore hybrids gas and electric vehicles. You will learn how they work and what types of hybrids there are. I will discuss their efficiency as well as costs and benefits — and even ways hybrids could be improved.

What Is a Hybrid?

A hybrid vehicle is one that can be operated on two or more sources of power or propulsion — for example, an internal combustion engine fueled by gasoline or diesel fuel and an electric motor powered by electricity generated by the vehicle itself (Figure 3.1). Propulsion systems in hybrids may operate separately or together: for example the electric motor can operate by itself or in conjunction with the gasoline engine to boost propulsion. The two means of propulsion are controlled by onboard computers that orchestrate operations to decrease fuel consumption and emissions.

Fig. 3.1: *Hybrids like the Toyota Prius are becoming widely available. These vehicles are powered by a gas engine and an electric motor supplied with electricity generated onboard.*

Hybrid vehicles appear to be a new invention, but the fact is, they've been around for a long time. Many readers may have unknowingly ridden in or even owned a hybrid at some point in their lives. If, for instance, you have ever driven a Moped, you've been on a hybrid vehicle. Mopeds are motorized pedal bikes. The operator starts off peddling, then switches to a gas engine once moving. I rode one in Europe in 1965 while studying abroad.

Unbeknownst to most of us, all diesel locomotives are hybrids. Although they are powered by diesel fuel, the diesel engine is there to run an electric generator. The generator, in turn, sends its electricity to an electric motor that drives the wheels of the locomotive. Technically, then, locomotives are diesel-electric hybrids.

Buses in Seattle are diesel-electric hybrids. They draw electricity from a spiderweb of overhead wires most of the time, but can also operate on diesel when traveling through parts of the city that lack the overhead wires. Large trucks used in mining operations are also frequently diesel-electric hybrids. All submarines are hybrids, too. Older subs, for example, are powered by diesel engines while most modern subs are powered by nuclear reactors. Both nuclear reactors and diesel engines produce electricity that's used to power the massive electric motors that turn the subs' propellers.

You may have been in a hybrid taxi. They're out there. Some taxicab fleets, like those in Vancouver, BC, contain hybrids. There are a few in Denver and other major cities as well. And some car rental agencies now offer hybrids. Maybe you've traveled in a hybrid owned by a friend or business associate.

Most hybrid cars and trucks are equipped with gasoline-powered internal combustion engines, although diesel can also fuel a hybrid. In fact, all three major US auto manufacturers developed extremely fuel-efficient diesel hybrids during the Clinton administration. (These prototypes achieved 70–80 mpg!) The French car manufacturer PSA Peugeot Citroën is at this writing developing two diesel-electric hybrid cars.

How Does a Hybrid Work?

Hybrids operate similarly to gasoline-powered cars and trucks, except for the addition of the electric motor, their second source of propulsion. To understand how a hybrid works, then, you first need to understand how a gas-powered vehicle works. (If you already understand the internal combustion engine, you may want to skip this section.)

A gasoline-powered car contains an internal combustion engine (ICE) that runs on gasoline, a mixture of medium-length hydrocarbons extracted from crude oil (petroleum) in the world's refineries. Gasoline is stored in a metal fuel tank, attached to the undercarriage of the car. It is pumped from the tank to the engine by the fuel pump. Fuel enters the cylinders of the ICE, where it is ignited. Ignition causes an explosion in the cylinders that drives the pistons down. (To view an animation of this, visit auto.howstuffworks.com/engine1.htm.) The pistons are attached to the crankshaft. As they are forced downward, they force the crankshaft to turn. (See auto.howstuffworks.com/engine2.htm for an animation.) The crankshaft, in turn, is connected to gears in the transmission, which turn the driveshaft. The driveshaft turns the front or rear wheels, depending on the car or truck.

A hybrid car adds several components to the mix, two of the most important being an electric motor and a battery. The electric motor is located beside the gas engine. Its job is to assist the gas engine by providing power when needed. It sometimes operates on its own, too, depending on the design of the vehicle.

The battery stores surplus electricity and feeds it to the electric motor when needed. The battery is not charged externally, for example, when the car is plugged into an electrical outlet, as with an electric car. Instead, hybrid gas-electric vehicles make their own electricity, as explained shortly. There are three types of hybrids on the market today, each of which makes electricity differently.

Figure 3.2 shows the first type, which is known as a *series hybrid*. In this configuration, the gasoline engine powers an electrical generator, known as an alternator. Electricity from the alternator powers the electric motor. It, in turn, drives the wheels. Surplus electricity is stored in the battery and used as needed.

The battery feeds direct current (DC) electricity to the electric motor. For this to work, however, the DC electricity must be converted to AC electricity by an inverter. The inverter converts the DC electricity stored in the battery into the AC electricity used to power the electric motor that drives the wheels of the car. As illustrated, the electric motor is attached

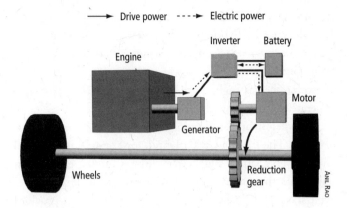

Fig. 3.2: *Series hybrids contain a small gas engine that powers a generator that makes electricity. Electricity is used to run the electric motor. Surplus can be stored in the battery. This design is extremely rare.*

to a reduction gear. It is connected to the front wheels via an axle. When the electric motor is running, the wheels turn.

To summarize, in a series hybrid, electricity is produced by a small gas engine. It drives the generator that produces electricity. The electricity can be fed to the batteries, where it is stored, or used immediately to power the electric motor, the only source of propulsion in this type of vehicle. Because the generator is used to make electricity, the gas engine and electric motor perform about the same amount of work.

Series hybrids are rare. The Coaster, a vehicle marketed in Japan, is one example. The Chevy Volt by GM is another member of this rare breed.

Far more common are *parallel hybrid* vehicles. As shown in Figure 3.3, a parallel hybrid consists of many of the same components as a series hybrid: a gas (or diesel) engine, an electric motor, a battery, and an inverter. In parallel hybrids, however, the electric motor and gas engines are arranged so that they both power the front wheels. As illustrated in Figure 3.3, they are linked to the front wheels via a common (shared) reduction gear. In a parallel hybrid, energy flows to the front wheels from either the gas engine or the electric motor — or simultaneously from both, depending on power requirements. In this design, however, the gas engine is the main energy source. That is, it operates more frequently than the electric motor.

Parallel hybrid technology has been the mainstay of Honda. The Honda Civic Hybrid, for example, is a parallel hybrid. The Saturn VUE Green Line is another example. In these cars, the gas engine is said to be the default. That is, when you start the vehicle, the gas engine starts up and provides motive power. However, when the gas pedal is pressed to climb a hill or pass another car, the electric motor kicks in. It provides additional thrust. These cars cannot operate on electricity only.

In a parallel hybrid, electricity is generated in two ways. One is by the alternator — just like in the series hybrid. The second source is the electric motor. Under certain conditions, the electric motor can generate electricity. That's why the electric motor is

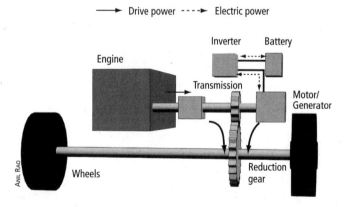

Fig. 3.3: *Parallel hybrids rely on the electric motor and gas engine, which power the wheels. The electric motor also generates electricity under certain conditions, for example when braking occurs.*

labeled "motor/generator" in Figure 3.3. How does an electric motor, which consumes electricity, also generate electricity?

Electric motors can become electric generators of electricity when spun faster than their normal operating speed. This occurs when the vehicle coasts down a hill or during braking. When this happens, electric motors produce AC electricity. AC electricity from the generator is converted to DC electricity and stored in the battery. It can then be used to power the electric motor when more power is needed, such as when climbing a hill or passing another car. This type of vehicle, however, can never run solely on electricity.

The third type of hybrid is a *series/parallel hybrid*. As its name implies, this design incorporates features of series and parallel systems. It taps into the battery-charging features of these two designs to reap the benefits of each. One of the main benefits of this unique marriage is higher gas mileage.

Figure 3.4 shows the main components of the series/parallel configuration. As you can see, it contains a gas engine, an

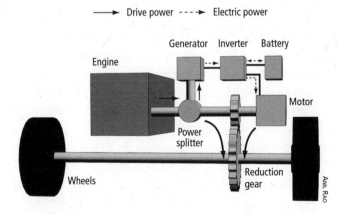

Fig. 3.4: *Series/parallel hybrids incorporate features of series and parallel designs, notably a generator and regenerative braking. They operate a bit more efficiently as a result.*

electric motor, and a battery. It also contains a generator, like the series hybrid, and a power split device, discussed below.

In a series/parallel hybrid design, a car or truck can operate on electricity only, gas only, or both, depending on the need. (It's all controlled seamlessly by an onboard computer.) For example, the Toyota Prius incorporates series/parallel hybrid technology. When the car starts out, it runs on electricity. Electric is the default. Electricity flows from the battery bank to the electric motor when the driver steps on the accelerator. This spins the front wheels and provides quiet, gas-free operation. My Prius can travel up to about 13–17 mph on electricity on a flat road. At higher speeds, the gas engine kicks in. From that point on, the car's computer regulates the operation of the gas and electric motors. When the Prius is cruising along the highway, the gas engine may provide the motive force. But when it's climbing a hill or passing another vehicle, both the gas engine and the electric motor operate. When it's coasting down a hill, both shut off.

How Regenerative Braking Works

To understand what's happening when a motor goes into "generator mode," remember that the electric motor actually acts like a generator the moment it begins to spin. If a generator is spinning *against a load,* it acts as a motor. Which effect — motor vs. generator — predominates determines whether we call it a motor or generator. Thus, if electricity is moving from the battery, we call it a motor. However, if the car is slowing down, the motor acts like a generator. The voltage that arises as a result of the "generator effect" can exceed that due to the "motor effect," and suddenly, the current moves in the opposite direction — to the battery.

In series/parallel hybrids, electricity is generated from two sources. As in a series hybrid, the generator produces electricity from the gas engine. However, as in a parallel hybrid, the electric motor/generator also produces electricity, for example during braking. That is, the electric motor becomes a generator when the vehicle slows down. This feature, known as *regenerative braking*, is described more fully in the accompanying sidebar.

The power splitter divides energy from the engine, according to the car's needs. That is, it controls the flow of energy to the wheels and the battery according to the car's needs and the requirements of the propulsion system. The series/electric system is found in the Estima Hybrid sold in Japan and the Toyota Prius, as noted earlier.

Not All Hybrids are Created Equal

As you can see, not all hybrids are created equal. In fact, hybrids exist on a continuum from "light" or "mild hybrids" with only a few gas-saving/pollution-reducing features, to "full hybrids" that offer the most fuel savings and pollution reduction. Even greater savings are available in the plug-in hybrids, discussed in the next chapter.

Table 3.1 lists hybrid features on a continuum. The first is the idle-off capability. Like the switch that turns the refrigerator light bulb off when you close the door, this feature allows a vehicle to turn off its gasoline engine when stopped at a stop light or stop sign, thereby saving fuel. What's amazing is that the engine turns back on in a fraction of a second. When you are ready to go, step on the accelerator, and the car starts moving instantly.

Buyer beware, though: some automakers have included the idle-off capability in some of their trucks and labeled them as hybrids. While this feature does increase fuel mileage, it doesn't make a car or a truck a hybrid.

Table 3.1
The Hybrid Feature Continuum

Idle-off capability (mild hybrid)

Regenerative braking (mild hybrid)

Power assist and engine downsizing (mild hybrid)

Electric drive only (full hybrid)

Extended electric ranges (plug-in hybrid)

The second feature on the continuum is regenerative braking. It is found in many mild hybrids, like the Honda Civic. The point of regenerative breaking is to capture energy that is usually lost to the atmosphere.

The energy associated with a car in motion is known as kinetic energy. The faster a car travels, the more kinetic energy it possesses. When a car slows down, kinetic energy is lost. Hybrids with regenerative braking are able to capture some of this energy during deceleration.

In a conventional gas- or diesel-powered vehicle, mechanical brakes slow the vehicle down. This is achieved by the brake pads clamping down on the rotor, a metal plate attached to the axles of the car. Friction created as the brake pads close down on the rotor slows the vehicle, then brings it to a complete stop. The kinetic energy (energy of motion) is converted into heat — hot brakes. The heat dissipates into the atmosphere, the kinetic energy being wasted.

With regenerative braking, hybrids rely on the electric motor to help slow a vehicle down. This occurs by converting the electric motor from *propulsion mode* to *generator mode*. During this conversion, explained earlier, the electric motor becomes an electric generator. As the car slows down, the kinetic energy is converted to electrical energy, which is stored in a battery.

The generator therefore recovers a portion of the kinetic energy that would otherwise have been lost. The kinetic energy stored as electricity in the battery pack can be used later, for example to speed up a car or start a series/parallel hybrid.

Again, however, beware of some automakers' claims. A vehicle's electric motor and battery have to be large enough to efficiently capture and store the generated electricity; otherwise there will be no real fuel economy. Some automakers claim that their gas-powered vehicles are equipped with regenerative braking when in reality these vehicles are only equipped with *integrated starter-generators* (starter motors that can generate electricity). They do not recover enough energy to significantly reduce fuel use or even power the vehicle. The main savings in such vehicles come from the idle-off capability.

The next feature of a hybrid is the power assist mode. It is, quite simply, the combination of a gas engine and an electric motor to create more power. This feature is found in parallel and series/parallel hybrids. Interestingly, the combination of gas and electric power results in a significant downsizing of the gas engine. My Toyota Prius (Gen II), for instance, came with a tiny 76-horsepower gas engine and a 67-horsepower electric motor. Those readers who understand engines may be put off by these numbers. Don't be. Manufacturers like Toyota can reduce the size of the gas engine without compromising the performance of a vehicle because the electric motor supplements the seemingly undersized gas engine, providing additional power when needed. If designed correctly, a hybrid can achieve the same power and performance as a conventional vehicle with much better fuel economy.

Vehicles containing idle-off, regenerative braking, and power assist are true hybrids, although they are considered mild hybrids by many. The Honda Insight and the Honda Civic fit into this group, even though the fuel economy of Honda's first version of the Insight was unsurpassed. (It got over 60 mpg on the highway!) The first version of the Insight was a lightweight two-seater commuter car (this version is no longer produced; in 2009, Honda came out with a new, four-seater Insight with a fuel economy similar to that of Toyota's Prius).

The next step up on the hybrid ladder is the capability of a vehicle to operate in electric-only mode. Vehicles with this feature are known as "full hybrids." Included in this small, rather elite group are the Toyota Prius and the Ford Escape Hybrid.

As noted earlier, the Prius generally starts on electricity — very quietly, I might add — and operates on electricity at low speeds. The gas engine kicks in as the car accelerates and operates primarily when it is most efficient — at higher speeds. (The only exception to the electric start of the Prius occurs on cold days. In such situations, the gas engine in series/parallel hybrids may come on to warm up the catalytic converter so that it's operational when needed. The catalytic converter breaks down unburned hydrocarbons and converts carbon monoxide to carbon dioxide. It has to be sufficiently hot to do this, so it may need to run for a bit to prepare for any unburned hydrocarbons sent to it from the engine.)

The final level of hybridization, discussed in detail in the next chapter, is the plug-in hybrid. Plug-in hybrids contain much larger battery packs and are programmed to allow the operator to drive in the electric-only mode for much longer distances — 20–60 miles — depending on the size of the battery pack. They also operate on electricity at higher speeds. Once the vehicle exceeds its electric-only range, that is, depletes its battery pack, it automatically switches to the gas-hybrid mode. This dramatically improves fuel economy.

Are Hybrids a Smart Economic Investment?

Although you may be interested in buying a hybrid to reduce your environmental impact, including your carbon footprint, you may wonder if the vehicle makes sense economically. Will the savings in gas offset the slightly higher cost of the vehicle?

The answer to this question depends on several factors. The most important are (1) the price of gas, (2) the fuel economy of the hybrid, (3) the fuel economy of the car you would be driving were you not to switch to a hybrid, (4) your driving habits, and (5) the cost of the vehicle. When considering switching to a hybrid or buying a hybrid instead of an ordinary gas-powered vehicle, you'll need to run the math. Fortunately, the calculations are pretty simple, as you will see shortly.

The decision to switch to a hybrid may make sense if gas prices are high and you purchase one of the less expensive hybrids, for example the Toyota Prius or Honda Civic. The Ford Escape Hybrid is rather pricey in comparison, costing nearly 30% more than the Prius or Civic hybrids (Figure 3.5). The decision would make economic sense if gas prices were high *and* the fuel economy of the vehicle you'd be driving instead of your hybrid was low, for example half to one third the fuel economy of a new hybrid.

Driving style is important, too. As noted in Chapter 1, how you drive a car makes a world of difference when it comes to

FORD MOTOR COMPANY

Fig. 3.5: *The Ford Escape Hybrid is an SUV series/parallel hybrid that uses technology licensed from Toyota.*

fuel economy — even with a hybrid. A heavy foot, aggressive driving in traffic, and other fuelish driving habits significantly reduce fuel economy and cost you dearly at the pump. If you don't modify your driving habits, you will not achieve the EPA's estimated fuel economy rating for your hybrid, although you may get better mileage than with your old gas guzzler. I recently drove my 2004 Toyota Prius 900 miles from my home in Evergreen, Colorado, to The Evergreen Institute's Center for Renewable Energy and Green Building in east-central Missouri. I kept the

Is a Hybrid Worth the Additional Cost?

Calculating the costs and benefits of a hybrid vehicle can be tricky. Consider this example: The Toyota Camry Hybrid has a 4-cylinder engine, but if you take into account the accessories that come with it and the fact that it performs like a V-6 Camry, it is best to compare the Camry Hybrid to a V-6 Camry LE. When you do, you'll find that the difference between the two cars' base prices is only about $1,400. At this writing, you would pay $23,640 for the Camry V-6 LE and $25,000 for the Camry Hybrid.

The Camry Hybrid's fuel economy is 33 mpg city/34 mpg highway; the Camry V-6 gets 21 mpg city/31 mpg highway. To compare the two, you can split the difference between city and highway mileage. The combined fuel economy of the V-6 Camry LE is therefore 25 mpg. The combined fuel efficiency for the hybrid is 33.5 mpg.

If you drive 15,000 miles a year, the hybrid version will consume about 450 gallons per year, while the non-hybrid V-6 LE will burn 635 gallons per year. At $3.20 a gallon, the hybrid will save about $550 a year. As a result, it will take about three years to recoup the higher up-front cost of the hybrid. ☛

speed at 60–65 mph and got 52 mpg. On another trip, I kept the mileage right at 60 mph and got 55.7 mpg. On a similar trip to Moab, Utah, I started out driving at 60, then toward the end of the trip, as we grew weary of the road, I increased the speed to 70–75 mph. I was amazed to see the gas mileage drop from around 50 mph to 43 mpg. (Air drag is … well … such a … drag!)

To calculate the costs of total ownership, be sure to factor in tax incentives such as state or federal income tax credits. Use that figure to compare the initial cost of the hybrid to that of

Another way of looking at this is to calculate the return on investment. Return on investment is 1 divided by the payback. A three-year payback, therefore, is the same as a 33% rate of return on your investment. (If only my stock portfolio performed half or a third as well!)

Consider another example: At this writing, the price of a 4-cylinder Honda Civic ranges from $14,810–$29,500, depending on the accessories and features. The Civic Hybrid sells for $22,600. The Civic Hybrid is rated 40 mpg city/45 mpg highway. The non-hybrid Civic is rated 26 mpg city/34 mpg highway.

To calculate the return on investment, we will use the Civic EX with a MSRP of $19,510. This car's combined mileage is 29 mpg. The Civic Hybrid, on the other hand, sells for $22,600 and gets 42 mpg in combined city and highway driving. The difference in cost is a little over $3,090, while the fuel economy difference is 13 mpg.

If gas is selling for $3.20 a gallon and you travel 15,000 miles a year, it will take a little over six years to pay for the higher price. While that may seem like a long time, it is actually a 16% return on your investment. ■

other vehicles. Once you've calculated initial costs, you can calculate the cost of gasoline for both cars, based on the number of miles you expect to drive each year and the cost of gasoline. Longer-term factors to consider are cumulative savings on brake pad replacements and oil changes. Through regenerative braking, the Prius is easier on brake pads than regular cars are, so they last longer. And oil changes aren't needed as often because the gas engine in hybrids like the Prius runs less often. So you can expect to spend less on these things.

Your calculations will very likely indicate that a hybrid will be cheaper to operate over the long haul, say three to five years, than a comparable car or a larger, fuel-inefficient SUV or mid-sized sedan. However, a hybrid will most likely cost more than a relatively energy-efficient car like the tiny Chevy Aveo or the slightly larger Chevy Cobalt — both dandy little cars.

However, don't forget that gas prices will very likely rise. Rising costs will make a hybrid more attractive economically. And, of course, not all decisions are based on money. The decision to buy that new fishing boat or new carpet wasn't made to save money. The same goes for a hybrid. You may want to drive one just because it is the right thing to do. It helps reduce gas consumption, decrease our nation's dependence on foreign oil, and dramatically reduce auto emissions, especially carbon dioxide. A few extra thousand dollars spent on a hybrid is a contribution you can make to creating a clean, healthy future for your children and the millions of species that share this planet with us. It also helps build a market for hybrids, which could eventually help lower prices. For a comparison of costs on some vehicles, see the accompanying sidebar.

How Could Hybrids be Improved?

Hybrids are great vehicles, with much promise, but they could achieve better fuel economy with a few changes. Reducing body weight is one of the most significant changes designed to improve gas mileage, especially in the city with all its annoying stop-and-go driving. If manufacturers used lighter-weight, but equally strong, composites (high-tech plastics with carbon fibers) to build the bodies, mileage could be boosted dramatically.

Fuel economy could also be improved if gas engines in hybrids were replaced with diesel engines. Although diesel engines cost more, they are significantly more efficient and more durable (longer-lasting) than gasoline internal combustion engines. They also require less maintenance, which helps to make up for the higher cost. As noted earlier, all three major US automakers have developed prototype diesel hybrids with a fuel economy ranging from 72–80 mpg (Table 3.2). (To learn

Table 3.2 Diesel Hybrids			
	GM Precept	**Ford Prodigy**	**Daimler Chrysler ESX3**
Fuel economy (mpg)	80	72	72
Engine	1.3 liter 3-cylinder diesel	1.2 liter 4-cylinder diesel	1.5 liter 3-cylinder diesel
Lightweight material	Aluminum	Aluminum	Thermoplastics
Coefficient of drag	0.163	0.199	0.22
Weight (pounds)	2,593	2,387	2,250
Battery	Nickel metal hydride or lithium polymer	Nickel metal hydride	Nickel metal hydride
Acceleration time (0 to 60)	11.5 seconds	12 seconds	11 seconds

more about these vehicles, visit autospeed.com/cms/A_0647/ article.html.)

If manufacturers combined these two features — diesel engines with lighter-weight bodies — fuel economy would climb even more. Lightweight diesel hybrid vehicles could easily get 100 mpg or more! If owners could fuel their cars with biodiesel or straight vegetable oil, this option would rocket to the top of the green car chart.

Another technology that could dramatically improve fuel mileage is the plug-in hybrid. Unlike conventional hybrids, which generate their own electricity, plug-in hybrids are charged externally — either from a 120- or a 240-volt electrical outlet. These vehicles operate as electric-only vehicles for trips up to around 60 miles, after which time they convert to the hybrid gas-and-electric mode, as noted earlier. This is made possible by incorporating larger, higher-capacity battery packs (lithium ion batteries) and larger and more powerful electric motors. The use of high-capacity lithium ion batteries increases the distance a plug-in hybrid can travel to at least 60 miles in the electric-only mode. Although that may not sound like much, 90% of all Americans drive less than 60 miles per day. Many Americans could drive their plug-in hybrids to and from work each day on electricity only.

Plug-in electric cars are not an unrealistic dream. They currently exist. In fact, several companies, like CalCars and Hybrids Plus, and a number of Toyota dealerships will gladly convert a conventional Prius (Gen II and Gen III) hybrid to a plug-in vehicle for about $10,000. This conversion increases the effective fuel economy of the Prius to 100–125 mpg. I'll explain the reasons for that in the next chapter.

Even though most of the electricity generated in North America comes from coal-fired power plants, plug-in hybrids are still cleaner than conventional gasoline- or diesel-powered vehicles. They could be even cleaner, however, if the electricity were generated from clean, renewable sources, such as solar or wind power. One of greenest cars I can imagine would be a diesel-powered plug-in hybrid made from lightweight composites made from chemicals derived from plants and operated on straight vegetable oil, biodiesel, or biomethane!

What to Look for When Buying a Hybrid

When shopping for a hybrid, beware of false or misleading claims. Remember, as pointed out earlier, not all hybrids are created equal. Be sure to look for the features listed in Table 3.1: idle-off capability, regenerative braking, power assist, and engine downsizing. The electric drive-only option, found in series/hybrid models, is, in my view, highly desirable, as cars with this option get the best fuel mileage of all hybrids. If you are interested in buying the greenest hybrid on the market, choose a series/parallel hybrid, unless, of course, series/parallel plug-in hybrids are available.

When shopping, it is also a good idea to check out the level of insulation in the cab, as well as the drag coefficient and dash meters and gauges. Why worry about such matters?

Insulation in a car reduces heat gain in the summer and heat loss in the winter. This, in turn, reduces the work required of the heater and the air conditioning system. Both can be operated less often or at lower settings. Reducing the run time and the load of the heater, air conditioner, and fans reduces electrical use, leaving more electricity for the electric motor. This, in turn, can boost fuel mileage.

Drag is created by friction as a car or truck moves through air. Reducing drag improves fuel economy. Interestingly, my favorite hybrid, the Toyota Prius (Gen II), has the lowest drag coefficient of any commercial car currently manufactured by a major automaker. The only car with a lower drag coefficient can be found in museums. It's the EV1, an electric car produced by GM, that was mothballed after five years, much to the dismay of those who had leased them (for the story on this tragic occurrence, see the video *Who Killed the Electric Car?*).

While you are shopping, be sure to check out gauges, especially the fuel economy gauge. It should provide instantaneous *and* cumulative fuel mileage feedback. The Gen II Prius came with a large screen in the middle of the dash. It was fairly easy to read at night but difficult to read during the day when the sun was out. The Gen III placed this gauge alongside the conventional digital displays right in front of the driver. Honda's gauges are located beside the other gauges directly in front of the driver, but are smaller and more difficult to read. Whatever you do, when you buy your first hybrid, be careful not to become obsessed with watching the fuel mileage gauge. I nearly ran a stop light at a busy intersection because I was intent on watching my Prius' fuel mileage. (Several other hybrid owners have admitted to the same problem.)

As with any car purchase, be sure to check out roominess, comfort, and drivability. Toyota's Prius not only gets better gas mileage than the Honda Civic Hybrid, but it is a much larger car. I've traveled cross-country with four sizeable men and suitcases without a hitch or complaint — except about my slow driving.

If you really want to green your wheels, you may want to hold off on your purchase until the plug-in hybrids or the Chevy Volt hits the market. According to my sources, Toyota is slated to release a plug-in hybrid Prius in 2010 for fleets (that is, company and government fleets). They will introduce the car for the rest of us in 2011. In 2009, they released a new, slightly larger version of the Prius. It provides additional features, one of the most impressive of which is a roof-mounted solar panel (it's mounted flush with the roof). The integrated solar panel generates electricity to run a fan to cool the car when it is parked in the sun on a hot summer day, reducing work required by the air conditioning system.

When shopping for a hybrid, be sure to check out various websites, like eHow's How to Buy a Hybrid at ehow.com/how_107420_buy-hybrid-car.html. As with any car purchase, be sure to test-drive different manufacturers' offerings. Don't be shocked, though, as I was the first time I test drove a Civic hybrid when the engine cut off after I came to a halt at a stop light. That's the idle-off feature, as noted earlier, which is designed to save gas. The car will start up immediately when you step on the gas pedal.

Would a Fuel-Efficient Car Perform as Well?

If a hybrid is too expensive, you may want to check out conventional fuel-efficient vehicles, for example Chevy's Aveo and Cobalt or Honda's Fit and Civic. Be sure to check out Toyota's Corolla and Yaris and Kia's Rio. You may even want to look at the Mini Cooper Clubman and Hardtop or Volkswagen's Jetta. Hyundai, Saturn, and Mazda also produce similar high-mileage vehicles.

When searching for an energy-efficient car, you may want to shop online first — to find what's best for you. You can compare features, fuel economy, and prices at autobytel.com or other similar online services. Be sure to check out the EPA's fuel economy website at fueleconomy.gov.

Energy-efficient gasoline-powered cars tend to be small, but affordably priced. They're all classified as economy cars and could make great cars for commuting or for short trips — or even long trips with few passengers. Bear in mind, however, that even though these cars are efficient, they are not generally as fuel efficient as hybrids. City fuel economy will be in the 20s, highway mileage in the 30s. The Honda Fit, for example, gets 28 mpg city/34 mpg highway, considerably lower than similar hybrids. The Kia Rio is rated at 27 mpg city/32 mpg highway, the Toyota Yaris is at 29 mpg city/36 mpg highway. With careful driving, however, you could boost the mileage into the 40s.

The Pros and Cons of Hybrids

Like all things, hybrid vehicles have their pluses and minuses. As noted earlier, hybrids reduce gas consumption, which reduces the cost of operating a vehicle. Reducing fuel consumption also reduces environmental pollution. A hybrid will also save you time, as you'll spend less of your life at the fuel pump.

Hybrids are widely available. Moreover, there's a wide array of full-hybrid cars on the market today, ranging from compact to mid-sized sedans to family-sized SUVs. Whereas in 2004, a US car buyer had only three choices — the Honda Insight, Honda Civic, and Toyota Prius — the list has expanded considerably. There are about 20 hybrid vehicles available today. Every major automaker offers at least one hybrid, sometimes up to three. Toyota, manufacturer of the Prius, the most popular hybrid on the road today, also offers a hybrid SUV, the Highlander, and a hybrid mid-sized sedan, the Camry. Toyota's high-end Lexus brand offers three hybrids. Honda, on the other hand, sells only the hybrid Civic and the Insight, which was reintroduced in 2009. The Insight is an extremely fuel-efficient car that gets about 40 mpg city/43 mpg highway. Honda is also working on a hybrid version of the Fit, which is rumored to get 80 mpg.

The hybrids I have driven perform admirably, even in mountainous terrain. My Prius takes to the mountains like a jackrabbit. It handles snow and ice with ease. (I always use the best snow tires I can afford, Firestone Blizzaks.) Hybrids drive like any other car, which is another check in the plus column. The onboard computer manages the dual power sources and provides a seamless driving experience. The only time you're aware you're driving a hybrid is when you examine your credit card statement at the end of the month and see fewer charges for gasoline.

In the early years, federal and state tax incentives were available to individuals buying hybrid vehicles, which lowered their initial cost. While federal incentives have been phased out, they could be resurrected under a president and congress attuned to the dire need for developing a green transportation system. (To check out federal incentives, visit the Department of Energy and EPA's website at fueleconomy.gov/Feg/tax_ hybrid.shtml.) Check with an accountant to see if your state offers rebates or tax incentives for the purchase of a hybrid. Remember, too, that many states allow hybrid drivers to use high-occupancy lanes (HOVs) — shaving time off the morning and evening commute. The lanes are typically reserved for vehicles with two or more passengers.

On the downside, hybrids cost more than similar gas-powered cars. That's because they are fitted with a lot more equipment, like costly batteries and electric motors. The Prius also comes with a plastic undercarriage screen that reduces aerodynamic drag along the underbelly of the vehicle. There is also a compartment that receives exhaust when the gas engine first starts up. Fumes are fed back through the catalytic converter once it has heated up. These and a host of other components add to the sticker price.

Mild hybrids cost less than full hybrids, but they don't deliver the fuel savings and emission reductions of their full-hybrid cousins.

Conclusion

As oil supplies dwindle, gas-electric hybrids will very likely become more popular. If you are thinking about buying a hybrid, now's a good time. I believe that conventional gas-electric hybrids will be around for a long time. However, as gas prices rise, expect to see more and more plug-in hybrids and all-electric vehicles. Electric cars with conventional battery technology could satisfy the needs of hundreds of millions of commuters the world over. Electric cars (discussed in Chapter 5) equipped with high-capacity batteries could provide more versatility, serving as commuter cars and medium-distance vehicles.

The plug-in hybrid could serve as a commuter car and a medium- and long-distance vehicle, providing even greater flexibility. When fueled by biodiesel or other renewable fuels, plug-in hybrids could become a major ally in our fight against global warming, acid rain, and urban air pollution. This option is discussed in Chapter 4.

CHAPTER 4

PLUG-IN HYBRID ELECTRICS

I magine driving a car that gets the equivalent of 100–125 mpg. Imagine, too, that during your 60-mile-per-day commute to work this car runs entirely on electricity. Imagine, as well, that this vehicle costs only 2–3 cents per mile to operate, compared to 12–15 cents per mile for a conventional 22-mile-per-gallon SUV. The vehicle would be even cheaper to run than an energy-efficient gasoline-electric hybrid, which costs about 6 cents per mile. If you've got any imagination left, imagine that this energy-efficient vehicle drives on longer trips, even cross-country. And, if that's not enough, imagine your new vehicle provides all these benefits while dramatically reducing emissions, including that pernicious greenhouse gas, carbon dioxide.

The vehicle I'm talking about is not a fantasy. It is a plug-in hybrid.

If several leading auto manufacturers are successful in their efforts — and there's every reason to believe that they will be — this dream could very soon become a reality on roadways the world over. But just what is a plug-in hybrid electric vehicle? How does it differ from conventional hybrids? When will they become available? What are the pros and cons of this exciting new option?

What is a Plug-in Hybrid?

Plug-in hybrids, or more correctly, plug-in hybrid electric vehicles, are very similar to the hybrid gasoline (or diesel) electric vehicles discussed in Chapter 3. They also incorporate many features of the all-electric cars discussed in Chapter 5. Like conventional hybrids, plug-in hybrid electric vehicles, commonly referred to as *PHEVs* in a bit of verbal economy, contain a gas or diesel internal combustion engine, an electric motor, and a large battery pack.

PHEVs are the "logical technology extension" of existing gas-electric hybrids, according to Roger Duncan of Austin Energy, a group that spearheaded a nationwide campaign to persuade automakers to build affordable plug-in hybrids. PHEVs are considered a logical extension of existing hybrids because a PHEV's batteries can be recharged internally (as in a conventional hybrid) but also externally — when plugged into electric power sources, for example an electrical outlet in your garage (Figure 4.1).

One of the leading proponents of PHEVs is CalCars, a nonprofit organization based in Palo Alto, California. This group consists of entrepreneurs, engineers, environmentalists, and citizens. They actively promote 100+mpg PHEVs. (For a discussion of theoretical fuel economy, see accompanying sidebar.) Calcars focuses its efforts on creating favorable public policy and technological innovation. It also has been pursuing ways to cultivate buyer demand, a component that is vital to the commercial success of PHEVs. By building demand among highly receptive markets — folks like you and me — this group is working to encourage automakers to manufacture 100+mpg "no-sacrifices" high-performance, clean hybrid cars.

PHEVs are not a brand-new invention, as noted in Chapter 3. The fact is, many Toyota Prius (Gen II and Gen III) owners have already converted their aerodynamic, fuel-efficient

Fig. 4.1: *Imagine plugging your car in at night after work, then driving the next day powered only by cheap, clean electricity. This car could make your dream a reality.*

gas-electric cars into PHEVs. What is more, PHEV conversions are occurring at such a rapid rate that it is no longer possible to track them. (You can check out the frequency of conversions on CalCar's website at calcars.org/news-archive.html.)

The California Cars Initiative converted their first Prius in September 2004, using a lead-acid battery pack — that is, conventional battery technology that's been around since the stone age ... well, almost that long. Lead-acid batteries are heavy and provide less storage capacity per unit weight than newer, more expensive innovations like nickel metal hydride and lithium ion batteries. They also charge considerably more slowly. In most PHEV Prius conversions today, the standard nickel metal hydride battery pack that was installed by the manufacturer is replaced by a lithium ion battery. One key reason for this switch is that lithium ion batteries are better able to handle

What Does 100 mpg Mean?

Auto manufacturers have, over the years, provided customers with fuel economy measurements of their vehicles in an effort to rate their cars. Unfortunately, no truly satisfactory measure of miles per gallon has resulted. That's because many factors come into play (e.g., highway vs. city, high vs. low altitude, etc.). Now, with the introduction of the PHEV, the confusion has only increased because of the type of "fuel" used in assessing mpg.

One group advocates a definition of *fuel equivalency*. That is, they talk about achieving fuel economy equivalent to 100 mpg or more in plug-in hybrid vehicles. They are not saying you're going to be burning gasoline at that rate. What they mean is that if you consider all of the fuel that goes into powering a PHEV, including gasoline and electricity, and then convert it to a common unit of measure, then convert it back to gallons, the fuel economy achieved by the huge amount of electricity and tiny amounts gasoline used to power the car would be equal to a fuel economy of 100 mpg or more.

Another group argues that this definition is uninformative and *misleading*. For example, what is the mpg rating of an electric vehicle (EV) that uses *no petroleum* but, rather, ☞

the deep discharging that occurs when a car runs on electricity for extended periods.

To convert a gas-electric hybrid Prius to a PHEV, installers add a plug and a battery charger. The battery charger converts the AC electricity from a standard electrical outlet to DC electricity. It is stored in the car's new battery. This conversion also requires a modification of the power control software of the car so it can operate at higher speeds and travel farther on electricity.

is charged from photovoltaic cells (PVs) or other renewable energy sources?

Attempts have been made to relate various forms of energy to an "equivalent gallon of gasoline," but some feel this is entirely impractical; they argue that a customer has a right to know, with no "double talk," how far a gallon of gasoline — with no electricity, downhill driving, 50 mph tailwinds or "Old Dobbin hitched to the reins"— will take them. As Dominic Crea has warned, "Failure to specify how far a vehicle can travel on a *specific* source of energy will render any measurement of fuel economy impossible and therefore meaningless. In fact, advertisers could claim that a PHEV version of the *Hummer* is more fuel efficient than a conventional *Prius* — and if you buy that, then take a look at these magic beans I have for sale." ∎

How far and how fast you can travel in the electricity-only mode depends on the driving conditions and topography. A friend of mine whose Toyota Prius (Gen II) was converted by a Toyota dealership reports that he can travel at 55–60 mph and go up to 60 miles on electricity on flat terrain before the car converts to the gas-electric mode.

Toyota's plug-in hybrid should be available in 2010, as this book hits the bookstores, but only for companies that purchase large vehicle fleets. They'll be available for sale to the general public in 2011, if all goes as planned.

PHEVs: A Technology Whose Time Has Come

PHEVs are not an off-in-the-distant-future technology like hydrogen vehicles that could take 30 years to reach commercial

The Father of PHEVs

Engineering professor Andy Frank has been working on PHEVs for over 30 years and has converted nine sedans and SUVs into PHEVs over the past decade with the help of his students at the Advanced Hybrid Research Center at the University of California, Davis. Their cars have won numerous awards in US Energy Department-sponsored "FutureTruck" competitions. Professor Frank is widely regarded as the "Father of the Plug-In Hybrid."

Owner Beware

When converting a Prius to PHEV, be sure the conversion is performed by an authorized party. Otherwise, a conversion can void the manufacturer's warranty. Contact your local Toyota dealer for advice or to hire them to do the conversion, if possible. (Not all Toyota dealerships are equipped and licensed for such conversions.)

production. PHEVs are ready for prime time in large part because they rely on existing technologies. No new advances, other than refinement of the battery packs, are required to make this dream an affordable reality.

The first commercially available plug-in hybrid — the Renault Elect'road (clever name, eh?) — became available in Europe in 2003. The Elect'road was a plug-in series hybrid version of the company's popular Kangoo. It had a 93-mile range, used nickel cadmium batteries, and came with a 500 cc (31 cubic inch) gas engine. The engine powered two high-voltage/high-output alternators. (Each alternator could supply up to 5.5 kW

Operating a PHEV

Under normal operating conditions, a PHEV with a full battery pack operates entirely on electricity. The driver simply charges the battery at home — usually at night — from a regular outlet at an equivalent cost of under $1 a gallon, or charges it at while at work — provided the employer or parking garage provides electrical outlets for charging EVs or PHEVs. When operated in this manner, then, a PHEV is an electric vehicle with a gas tank for backup. Should a driver need to travel longer distances, that is, beyond the range of the batteries, the car switches over to gas and electric mode, like a standard (non-plug-in) hybrid. In such instances, the gas engine takes over. It charges the batteries (through the generator), as described in the previous chapter on conventional gas-electric hybrids.

at 132 volts at 5,000 rpm.) The fuel tank of this vehicle held 2.6 gallons of gas. The passenger compartment was heated by electricity from the battery pack (which powered a resistance heater and blower fan) and also a conventional heater. (In conventional cars, heat is provided by engine coolant that passes through the gas engine.) Renault sold 500 vehicles, primarily in France, Norway, and the U.K.; they redesigned the vehicle in 2007.

Another pioneer in the PHEV market is China's BYD Auto. They released a hatchback PHEV in December 2008. This car is known as the F3DM PHEV-68 (PHEV-109km). As the name implies, it can travel 68 miles or 109 km on electricity. The car sells for 149,800 yuan ($22,000).

At first reluctant to venture into this realm, many other automakers are now pursuing PHEVs. From 2004 to 2006,

PHEV Terminology

In plug-in hybrids, the all-electric range is designated by a number following "PHEV-," for example, PHEV-*[miles]* or PHEV-*[kilometers]* km. This number is the distance the vehicle can travel on battery power alone. As an example, a PHEV-60 car could travel sixty miles using only electricity from a full battery pack. The internal combustion engine is not required during this time.

for example, Toyota, General Motors, Ford, DaimlerChrysler, Volkswagen, and two California startup companies, Fisker Automotive and Aptera Motors, all announced their intentions to introduce production PHEVs in the near future. Ford's PHEV-30 is currently being used in utility fleets and will be available for public purchase in 2012.

DaimlerChrysler has tested PHEV prototypes and has converted as many as forty 15-passenger Mercedes commercial vans into PHEVs. Some of these vehicles use nickel metal hydride batteries. Others are equipped with more advanced lithium ion batteries. Some have diesel engines; others are equipped with gasoline engines. This work is being carried out in cooperation with California's Electric Power Research Institute (EPRI), the South Coast Air Quality Management District and Southern California Edison.

General Motors has also announced that it will soon mass-produce two plug-in hybrid electrics, the Chevy Volt and the Saturn VUE. (With the demise of Saturn, though, this vehicle will probably never make it to market.) The Chevy Volt is a series hybrid sedan. It contains a small internal combustion engine that powers a 53-kW electric generator. The generator produces electricity that powers an electric motor that drives the wheels. Surplus electricity is sent to the batteries. The distribution of electricity is controlled by the electronic control unit, a technology found in all hybrids.

According to GM's website, the "Chevy Volt is designed to move more than 75 percent of America's daily commuters without a single drop of gas. For someone who drives fewer than 40 miles a day, the Chevy Volt will use zero gasoline and produce zero emissions." (Note that while this may be true *at the tailpipe*, the Volt is powered by electricity that has to come from somewhere; at the moment that means from power plants that produce pollution and emissions.) The gasoline-powered engine that drives a generator will provide electricity to power the electric motor when the car is driven beyond the 40-mile battery range. On long trips, then, a Volt with a fully charged battery pack would operate on electricity for the first 40 miles, then the gas engine would kick in, producing electricity to power the car for the rest of the trip. When the car starts with a full tank of gas, the gas engine could extend the Volt's range up to 640 miles, depending on terrain, road conditions, and driving speed. When the gas gauge hits empty, the driver would simply fill up and continue the trip. At the destination, the driver can recharge the car and operate it in electric-only mode for all trips under 40 miles.

The Chevy Volt incorporates a lithium ion battery pack. According to the manufacturer, recharging requires eight hours on 120-volt electricity and about three hours on 240 volts. What is more, the engine is designed to operate a number of different fuels, including E-85, diesel, hydrogen, and natural gas, making this the most versatile hybrid on the market.

If electricity is selling for 10 cents per kWh, the Volt would cost about 80 cents to $1.00 a day to drive 40 miles — on electricity. In contrast, a Toyota Prius that achieves 42 mpg would consume one gallon of gas for the same commute, which would cost $2–$4 in the US, depending on the price of fuel.

The Chevy Volt, designed to seat four, is about the size of a Honda Civic sedan — just two inches wider. The Volt is also two inches longer than the Prius (Gen II), but it carries less cargo — the Volt's cargo capacity is 10.4 cubic feet versus 14.4 in the Prius (Gen II). The Chevy Volt is scheduled to go into production in late 2010. To learn more about this vehicle, go to chevrolet.com/electriccar/.

The Pros and Cons of Plug-In Hybrids

It's hard to find much wrong with plug-in electric hybrids, other than cost. Although they perpetuate our dependence on oil, especially foreign oil, they do have the potential to dramatically reduce gas consumption. If you used a PHEV for local commuting with distances less than the car's electric-only range, you might never have to fill up with gas again. You could travel solely on electricity supplied in your own garage! Plus, if you needed to take a long trip, with PHEVs you wouldn't be constrained by the limited battery capacity of electric vehicles, discussed in the next chapter.

PHEVs are cheap to operate. According to Calcars.org, you can fill up your batteries for less than the equivalent of $1 a gallon. According to their conservative estimates, using the average US electricity rate of 9 cents per kilowatt-hour (kWh) and assuming the car averages 25 mpg, 30 miles of electric driving would cost 81 cents. That's equivalent to 75 cents per gallon. More fuel-efficient PHEVs will perform even better.

If PHEVs run mostly on electricity rather than gasoline, won't countries need to build new power plants to meet increased demand? The answer is very likely no.

Because PHEVs are primarily charged at night, owners can charge their cars' batteries using nighttime electricity from local power plants. As a rule, most utilities power down their plants at night as a result of reduced demand. In many areas, power companies charge less for electricity generated at night. In fact, nighttime electric rates can be as low as 2–3 cents per kWh, which is the equivalent of 20–25 cents a gallon. "As PHEVs start to enter the marketplace," say the folks at calcars.org, "we'll see increasing support from electric utilities." They could even offer reduced nighttime rates to encourage owners to charge their vehicles during off-peak hours — that is, during times when demand is low.

PHEVs could help us utilize renewable energy resources more efficiently as well. In some areas, for example, in Texas, winds often blow at night. Without a means to store the surplus, utilities that generate electricity with wind turbines are unable to use the surplus power. Plug-in hybrids could consume this clean, renewable energy surplus.

What if coal remains the dominant source of electricity? Won't all the electricity produced by coal-fired power plants just make things worse? Won't it increase acid rain, global warming, and urban air pollution?

Although it is dead wrong to label an electric car "emission free" — there's always some emission at power plants — electric vehicles produce far less pollution when powered by electricity from a coal-fired power plant than conventional gas- and diesel-burning cars. According to Sheri Boschert, author of *Plug-in Hybrids: The Cars that Will Recharge America*, when powered by electricity produced by coal-fired power plants, PHEVs produce 42% less carbon dioxide than comparable gasoline-powered vehicles. (This statistic comes from the National Renewable Energy Laboratory in Golden, Colorado.)

A report by the Electric Power Research Institute and the National Resources Defense Council notes that PHEVs, unlike gasoline cars, could get cleaner as they get older because the electrical grid itself is (slowly) getting cleaner. New coal-fired power plants that replace retired facilities produce considerably

Gasoline and Electricity

One gallon of gasoline contains as much energy as about 36 kWh of electricity. (For the technically minded reader, it's the *thermal equivalent* of 36 kWh of electricity.) But that doesn't mean 36 kWh of electricity equals one gallon of forward motion. A typical internal combustion engine runs at about 25% efficiency. Diesels run significantly higher, at about 40%. An electric car, on the other hand, is about 65% efficient, due to the high conversion efficiency of its electric motor.

So what does this mean to you?

One gallon of gas will propel a conventional car about 25–30 miles. Thirty-six kilowatt-hours of electricity, which is equivalent to one gallon, will propel an electric car of similar size not quite three times farther — actually 2.6 times farther. In this comparison, then, 14 kWh of electricity has the propulsive power of a single gallon of gasoline. At 10 cents a kWh, a gallon's worth of fuel will cost $1.40.

less pollution. Furthermore, more and more electricity is being generated by clean, reliable, and renewable resources such as wind and solar energy.

Studies show that under a wide range of possible scenarios, PHEVs could dramatically reduce greenhouse gas emissions over the next four decades. Even if coal combustion continues to be a major source of electricity in the next two decades, replacing gasoline-burning autos, trucks, and SUVs with plug-in hybrids could very well improve air quality.

PHEVs make sense for individuals and businesses that want to help us end our addiction to oil, combat global warming, and address a host of other environmental issues that result from our dependence on this fossil fuel. Not only can automakers manufacture PHEVs now, using existing technology, but PHEVs can plug into *today's* infrastructure — the power grid. We won't have to create a whole new infrastructure to make this technology work, as would be necessary for hydrogen.

PHEVs will cost more than conventional hybrid vehicles — an estimated 10–20% more. Expect to pay $2,000–$3,000 more for a sedan and $5,000 more for an SUV, according to calcars.org. Incentives from state and national governments could help lower the costs, however. And don't forget, you'll also save a huge amount of time over the course of a year by avoiding visits to your local filling station.

Another possible advantage of plug-in electric hybrids, say proponents, is that these vehicles could become a backup source of power in times of shortage — for example, during a blackout. PHEVs could power a home if the grid goes down for a day or two due to an ice storm or thunderstorm. Surplus electricity stored in the batteries could be fed into a home, keeping it operating more or less normally. (Homeowners would need to prioritize their energy use, and power only critical loads like fridges, heater fans, boiler pumps, etc., during such times.)

Some utilities have also expressed an interest in occasionally buying surplus electricity from PHEV owners during peak demand — that is, at times when it is difficult for them to meet the demand of their customers (Figure 4.2). During peak demand, many utilities purchase power on the spot market from other providers. They often pay premium rates for this electricity, sometimes ten times more than their cost to generate electricity. Buying surplus electricity from local producers — that is, PHEV owners and individuals like you and me who have installed solar electric systems or small wind energy systems to power our homes and our vehicles — could save utilities a lot of money over the course of a year. This application, known

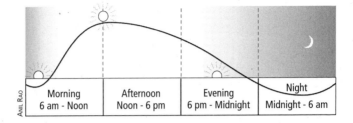

| Morning
6 am - Noon | Afternoon
Noon - 6 pm | Evening
6 pm - Midnight | Night
Midnight - 6 am |

Fig. 4.2: *Electrical demand rises during the day, peaking as human activities increase. Demand can sometimes exceed the capacity of a local utility's power plants, making it necessary to purchase power from outside sources. PHEVs can supply electricity to the grid during such times. They can also be charged during periods of reduced demand, helping to even out the load.*

as *vehicle-to-grid* (V2G) power, benefits the utility and all its rate payers.

PHEVs could also be a boon to people like me who live off-grid and whose renewable energy systems often produce surplus electricity. For many of us, once our batteries are full, the surplus is wasted. A PHEV in the garage could "absorb" this surplus. (I just checked: my PV system is producing a surplus of electricity right now, and it is only 11 AM. It's been producing surpluses almost every day of the year, except during the winter, when it is cloudier and electrical demand in my home and office is highest.)

In closing, PHEVs are good for both commuting and longer trips. They are likely to become the primary mode of transportation in the not-too-distant future. As such, PHEVS could become a short-term and long-term solution to many of the problems the US and a host of other countries are now facing. Modifications that permit them to run on clean, sustainably produced renewable fuels, such as biodiesel, biomethane, and ethanol, could ensure them a place in a sustainable transportation system.

If we're smart, PHEVs could help us achieve energy independence, replacing Middle East oil with homegrown American wind and solar energy. Improvements in batteries and the use of high-strength, lightweight composites could boost the range and fuel economy of PHEVs. It is not inconceivable that a strong, but lightweight PHEV could travel 200–300 miles on a single charge and could achieve a fuel economy equivalent to 150–200 mpg!

CHAPTER 5

THE ELECTRIC CAR

Because they offer so many benefits, electric cars and trucks, referred to as electric vehicles or EVs for short, will likely secure a place in a sustainable system of transportation. Although EVs aren't currently designed for long trips, they are ideal for the daily commute and running errands. That said, technological improvements could make some EVs excellent mid-range vehicles, allowing 200 miles of travel or more on a single charge.

In the future, then, families with two cars in their garages will very likely own an electric car for the daily commute, errands, and mid-range trips, and a plug-in gas- or diesel-electric hybrid for commuting as well as mid- and long-range trips. This chapter explores electric cars.

How Electric Cars Work

Figure 5.1 shows the anatomy of an electric car. Those who are familiar with vehicles powered by internal combustion and diesel engines will find the electric car to be a much simpler beast.

At the rear of the vehicle, there is often an onboard battery charger, commonly referred to as the charger. (In some EVs, battery chargers are located externally in the garage or at special charging stations.) The battery charger plugs into an electrical outlet, most often a 120-volt outlet. It converts alternating current (AC) electricity from the outlet into direct current (DC) electricity that is fed into and stored in the vehicle's batteries.

To start an electric car, the driver turns a switch or key. This activates the main contactor, shown in Figure 5.1. This device is a relay that, when closed, permits electricity to flow from the batteries to the controller. The controller is like the carburetor in a gasoline-powered car. It regulates the flow of fuel (in this case, electricity) to the motor. As indicated in the diagram, electricity flows from the batteries to the controller and then to the electric motor. The controller regulates the amount of electricity that flows to the electric motor. Just like the carburetor in a car, the controller is regulated by pressure applied by a driver's foot on the "gas" pedal. The harder the pedal is

Fig. 5.1

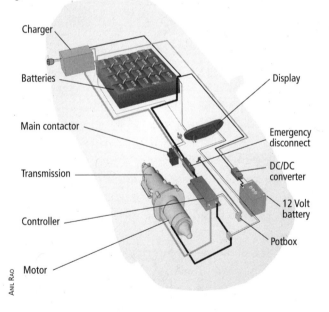

Charger
Batteries
Main contactor
Transmission
Controller
Motor
Display
Emergency disconnect
DC/DC converter
12 Volt battery
Potbox

ANIL RAO

pressed, the more electricity flows to the motor. Pressure from the accelerator pedal is, in turn, converted to an electric signal that is transmitted to the controller by yet another component, known as the potbox (because it is a potentiometer). As illustrated in Figure 5.1, a cable runs from the accelerator pedal to the potbox.

The electric motor is the workhorse of an EV. It converts electrical energy into mechanical energy that moves the vehicle forward. Motors can be AC or DC. In electric cars equipped with AC motors, the electric motor can perform two functions. It can power the car and, under certain conditions, can act as an electric generator. As you may recall from Chapter 3, a motor acts as a generator when it spins above its operating speed. This occurs when the motor is no longer in use, for example when the car is slowing down. This feature, known as regenerative braking, produces electricity that is sent to the batteries. It captures some of the kinetic energy (energy of movement) that would otherwise be lost as a vehicle slows or comes to a stop.

As shown in Figure 5.1, electric cars contain an emergency disconnect, which is wired into the system between the batteries and the controller. The emergency disconnect is a switch and a breaker, much like the circuit breakers in the breaker box of a home — though much bigger (Figure 5.2). It automatically disconnects the batteries from the rest of the vehicle's electronic and mechanical components if a short circuit occurs in the system. This protects wires and other components from potentially damaging high-current flow. It also can be manually switched off in case the vehicle needs to be serviced or is involved in an accident.

As shown in Figure 5.1, EVs also contain a single 12-volt battery located in the front of the vehicle. It provides electricity to power the gauges, radio, windshield wipers, lights, and other

DAN CHIRAS

Fig. 5.2: *This massive circuit breaker can be used to disconnect the batteries during an accident or when service is required on a vehicle. It also acts as an automatic circuit breaker in the event of a short circuit in the system.*

electronic components. This battery is recharged from the main battery pack, often referred to as the *traction battery pack*.

Since the main battery pack is typically wired to produce very high voltage (94 to over 300 volts), EVs must contain a DC to DC converter to charge the 12-volt battery. Located near the 12-volt battery, it reduces the incoming voltage from the traction batteries, thus protecting the battery. In a car with a 144-volt traction battery, for instance, the DC/DC converter lowers the voltage from 144 to 12 volts.

How Do You Get an Electric Car?

If you are thinking about switching to an electric vehicle, you have three options to get your tired bones behind the wheel of one of these clean, quiet, and powerful vehicles. First, you can purchase a used EV produced by a major auto manufacturer such as Toyota or Ford in the 1990s to early 2000s. They were produced in limited quantity and are rather difficult to find. Expect to pay a lot for one of these classics!

Or you can buy a brand new EV vehicle produced either by a major auto manufacturer like Nissan, or by a smaller company that manufactures only EVs.

Or, third, you can convert your car, truck, or van to electricity. In the next section, I'll examine all three options, starting with the easiest one, purchasing a new EV.

Buying a New EV

In September 1990, the California Air Resources Board mandated that, beginning in 1998, all major automakers must sell zero-emission vehicles in the state as part of an effort to reduce air pollution. The EV was the perfect candidate.

In January of 1990, at the Los Angeles Auto Show, the president of General Motors had already introduced an EV concept car, a two-seater, known as the Impact. Between 1996 and 1998 GM begrudgingly produced, sold, and leased 1,117 of this very popular EV1. Eight hundred of these sleek, aerodynamic vehicles were made available through three-year leases, which were quickly snatched up by eager drivers. Chrysler, Ford, GM, Honda, Nissan and Toyota also produced a limited numbers of electric vehicles for sale in California.

By all accounts, the EV1 and other EVs performed well and were extremely popular among those who purchased or leased them. Unfortunately, in 2003, when the leases expired on their EV1s, GM snatched the cars back from customers. In a storm of controversy, the company destroyed most of the cars, crushing them. This tragic event, chronicled in the movie *Who Killed the Electric Car*, has been attributed to several factors. First was the auto industry's successful challenge in federal court of

California's zero-emission vehicle mandate. The law having been struck down, the car was, at least in the eyes of the regulators and manufacturers, rendered obsolete. Second was a federal regulation requiring GM to produce and maintain spare parts for the EV1. If the car were no longer on the road, the company would not be required to comply with this mandate. Third was a successful oil and auto industry media campaign aimed at reducing public acceptance of EVs.

Honda, Nissan, and Toyota had also made their cars available by closed-end leases and took possession of them at the end of the lease period. Like GM, they crushed their vehicles, removing them from the road. Bowing to public pressure, however, Toyota reluctantly agreed to sell 200 of its RAV4 EVs to customers. Many of these cars are still on the road today, commanding a price higher than their original $40,000 sticker price.

Fortunately, the electric car has begun to make a comeback. Because of rising gas prices in the early to mid 2000s, concern over climate change, and customer demand, all major carmakers and many smaller companies have announced plans to produce and market EVs throughout the world, starting in 2010. Among the participants in this effort are Nissan, Toyota, VW, Daimler AG, GM, Renault SA, Peugeot-Citroën and Mitsubishi.

At the North American International Auto Show in Detroit in January 2009, for instance, Toyota announced plans to manufacture an all-electric commuter car by 2012. At the show, Toyota showcased an all-electric, battery-powered compact

Fig. 5.3: *Expect to see this and other all-electric commuter cars on the road in years to come.*

concept car, called the FT-EV. In 2010, Nissan released the Leaf, an electric sedan.

Some of the new EVs slated for release are sports cars — great commuter cars if you are financially well off. In 2008, for instance, Chrysler announced production of an all-electric Dodge sports car based on the Lotus Europa. Intended for sale in North America, it is a handsome car that will cost a small fortune. If an all-electric sports car is within your budget, consider the $100,000 Tesla Roadster, a sleek, attractive and sporty all-electric vehicle produced by California's Tesla Motors. The car goes from 0 to 60 in under 4 seconds, achieves a top speed of 125 mph and can travel up to 200 miles on a single charge of its lithium ion batteries. Fortunately, Tesla Motors has developed a more affordable electric car, the Model S, for mainstream buyers, too. (But it's still not cheap.)

Be sure to check out smaller producers like Zap as well. They manufacture and distribute electric cars, including very reasonably priced vans and small trucks. These vehicles are all fairly small and not intended for high speed or long distances (over 30–40 miles). Their vans and trucks, for instance, top out at 25 mph and cost about $15,000. The Xebra Zap car tops out at 40 mph and costs around $11,700. If you are looking for something sportier and faster, check out the three-wheel sporty Alias (Figure 5.4) featured on the front cover of this book. Its top speed is 105 mph, and it can travel 75+ miles on a single charge. The Alias is listed for about $35,000 in 2010. The company also manufactures electric scooters, bikes, and ATVs. You can see them at zapworld.com.

Another cool EV is the three-wheel, all-electric space-age Aptera. To see it, visit aptera.com.

Canadian buyers should check out Zenn Motor Company at zenncars.com. "Zenn" stands for "zero emission, no noise." I should point out that both claims are exaggerated. All electric

Fig. 5.4: *This sporty car can travel 75+ miles on a single charge and is relatively affordable.*

vehicles still involve air pollution at power plants unless they are run on electricity from renewable energy. Zenn Motor Company produces an attractive low-speed commuter car (its top speed is 25 mph) with a 30–50 mile range.

European car buyers should check out the Norwegian automaker, Think. They sell a compact EV car known as Think City (think.no). Also be sure to check out Mitsubishi's iMiEV Electric Car. Mitsubishi will be supplying iMiEV electric cars to French automakers Citroën and Peugeot. European buyers should also check out Renault. They're developing several electric vehicles.

Individuals interested in buying an EV can find a wealth of information on the web at sites like ElectricCars (electriccars.com/main.cfm) and EV World (evworld.co). The Electric Auto Association's newsletter, *Current EVents*, is a goldmine of useful information. It's packed with informative articles on the latest developments in electric cars.

Buying a Used EV

Buying a preowned EV is another option, though used EVs are few and far between. I can't emphasize enough how important it is for buyers to do their homework first. Know what you're looking for and looking at. Don't make rash decisions. Not all used EVs are created equal. In fact, there are significant differences. Because used EVs are rare, you might be tempted to buy the first one that comes along. Be patient. Shop carefully and wait for the perfect EV, even if it takes a year or two.

When buying a used EV, you have two options. First, you can purchase one manufactured by a major auto manufacturer, such as General Motors, Ford, or Toyota. A good example is Toyota's RAV4 EV. This small SUV is one of the best ever produced. Your second option is to buy an EV that an individual or a mechanic has converted to electricity. Let's start with the first option.

According to Shari Prange, who has written extensively on EVs in *Home Power* magazine, the only factory EVs you're likely to find are small electric pickup trucks, the Chevy S-10 and Ford Ranger, and Toyota's RAV4 I just mentioned. All were produced in limited quantity for a short period.

To locate an EV, go online. Check out Craigslist and eBay. It is important to remember that the only EV that's still supported by the manufacturer is the Toyota RAV4 EV. (The company faithfully provides parts should the vehicle need repair.) GM and Ford do not provide parts for their EVs. If something goes wrong with one of these vehicles, you'll need to find parts on the open market place. This may be possible, but it will be more challenging than ordering directly from the automaker.

Pay attention to the type of charging the vehicle requires. GM's and Toyota's EVs, for instance, require a special external charging station, which may be difficult to locate. The Ford Ranger is equipped with an onboard battery charger. However, it requires a special connector that won't plug into a standard receptacle. You'll need an adaptor to make it work. Because all these vehicles are in pretty high demand, expect to pay $30,000–$40,000 for them.

When shopping online, you may also encounter used EVs produced during the 1970s through the early 2000s by several smaller auto companies. These include the CitiCar, Comuta-Car, Tropica, Lectric Leopard, and Solectria Force. The first three — CitiCar, Comuta-Car, and Tropica — were built from scratch. That is, they were designed and built as EVs. The last two — Lectric Leopard and Solectria Force — were converted to EVs by the manufacturer from two gasoline-powered models, the Renault LeCar and Geo Metro, respectively. As with most other commercially produced EVs, the big problem with these cars is that they're no longer supported by the manufacturers. If you have problems, you'll have to purchase parts through private EV suppliers.

Another option is to purchase a gas-powered vehicle that has been converted to an EV by a mechanic or do-it-yourselfer. Be sure to ask why they are selling their car, and be sure to test-drive the vehicle first. EVs converted from conventional cars and trucks are more widely available than manufactured EVs.

When shopping, avoid cars that were cobbled together with parts from various suppliers. In such instances, good quality parts may be paired up with poorer quality components. These cars are more likely to have problems. Cars converted with universal kits (discussed below) are a better choice, provided the workmanship is good. Your best bet is a car converted with a custom EV conversion kit, a kit designed specifically for a given car — so long as the job was done right. If the car was converted by a qualified mechanic, so much the better. For advice on buying a used EV, I strongly recommend you read the series of articles published in *Home Power* magazine by Shari Prange and her husband Mike Brown. They were published in issues 89, 93, and 118. Be sure to also read the next section, on converting a car to electricity.

Converting a Car to Electricity

If you're a skilled and patient mechanic, you can convert your own car to electricity. Expect to invest at least 40–60 hours in the project. For best results, it is worth purchasing a kit that comes with compatible parts — an electric motor, controller, potbox, charger, emergency disconnect, adaptor, etc. — that will work together flawlessly. When shopping online for kits, you will encounter a number of companies. Be careful though: nine months after ordering a kit from a prominent California-based company, I still hadn't received two vital parts worth about $2,500 each. Check around and ask others who they've worked with recently, and contact the Better Business Bureau. Expect to pay $4,000–$10,000 for a kit.

EV conversion kits come in two varieties: universal and custom. Universal kits contain parts that have been carefully selected to work well together. They can be used in a number of similar cars and trucks. Custom kits contain parts that are selected for a particular vehicle, for example a Volkswagen Rabbit or Beetle.

Shocked by the high price of kits, some adventurous — some would say foolhardy — souls try to go it on their own. They purchase the parts separately, mixing and matching without sufficient attention to detail. Unfortunately, this path is fraught with danger. Many who embark on it end up with cars that don't last long or don't work well — or, worse yet, don't work at all. For best results, be sure to carefully match parts. Consult with experts and don't skimp on parts.

Technically, any vehicle can be converted to electricity, but for best results, you should probably choose a used lightweight vehicle for which there's a custom kit. Stay away from heavier vehicles such as vans or half-ton or three-quarter-ton pickups. Heavier vehicles require massive battery packs; because they are so heavy, you can't travel very far on a charge. For best results, select a car with a curb weight under 2,500 pounds, preferably under 2,000 pounds. (Curb weight is the weight of the car prior to conversion, without passengers.) The upper limit is 3,000 pounds. That doesn't leave many choices.

When converting a car to electricity, you'll have to first remove the gas engine, cooling system, exhaust system, starter, gas tank, fuel lines, and a few other components found in vehicles with internal combustion engines (ICEs) (Figure 5.5). You may even want to remove the transmission, for reasons described shortly. A copy of the vehicle's shop manual is vital when it comes to removing certain parts. You'll also need to carefully remove and properly dispose of all fluids, such as crankcase oil and engine coolant. Be sure to recycle parts, too.

Stripping an internal combustion engine car of its guts lightens it considerably. However, once the conversion is complete, most vehicles gain about 800 pounds — sometimes more. That's primarily due to the addition of the batteries, which weigh around 70–80 pounds each. (You will need to

Fig. 5.5: *The first step in converting a vehicle to electricity is to strip out the internal combustion engine and all related parts that won't be used in the EV.*

install 12–24 batteries, depending on the voltage of the system, which depends on the weight of the vehicle.)

Although lightweight cars like VW Beetles are ideal for conversion, smaller vehicles pose some problems. One of the most significant is that they have very little room for batteries. For best results, select a vehicle that is light but roomy, for example a VW Rabbit, Honda Civic, Nissan Sentra or Ford Escort, a lightweight pickup truck like the 1970s Datsun pickups or a medium-weight pickup like the Chevy S-10 or Ford Ranger. In 2010 and 2011, I converted a Chevy S-10 to electricity. Its curb weight is a little over 3,000 pounds (as of April 2011, I'm still waiting for parts from the supplier.)

You'll also need to select a car that's in good shape and mechanically sound. Most conversions are performed on manual transmission vehicles. Automatics present some serious problems because the engine has to keep running when the vehicle is stopped. For more on selecting a car for conversion, you may want to read Brown's and Prange's book, *Convert It!* It offers a lot of sound advice. Before you embark on an EV conversion, it's a very good idea to attend a workshop or two. Unfortunately, there aren't many places in the US that offer workshops. You can learn about my conversion at my website, evergreeninstitute.org.

Once you've purchased a car for an EV conversion and stripped out the internal combustion engine and associated parts, you'll need to install the electric motor. It is mounted in the center of the engine compartment and attached to the transmission (Figure 5.6). Other components, like the controller and main contactor, are then mounted around it. Most people who convert their cars to electricity install DC motors. Choose very carefully. There are not many DC motors that work well in a car.

To attach the electric motor to the transmission, you'll first need to attach an adaptor to the motor (Figure 5.7). Adaptors

DAN CHIRAS

Fig. 5.6: *The electric motor (right) is attached to the transmission outside the vehicle for ease of installation. The entire assembly is then lifted into place.*

are specially designed to connect the electric motor to the existing transmission. As Brown and Prange note, "All of the horsepower and torque needed to move the car down the road must pass through the adaptor on its way to the transmission and wheels." As a result, adaptors must be carefully designed and sturdily built. Fortunately, a good custom kit will contain the proper adaptor for your conversion. If you buy a kit from a supplier that sells high-quality kits, you should be okay. (Look for a company that's been in the business for a long time, and don't let them charge you until the parts ship.) Don't skimp on this component and don't try to make your own. You'll quite literally pay for a shoddy adaptor down the road, when the car grinds to a halt because the adaptor broke.

Once assembled, the adaptor/motor/transmission is lowered into place. A battery containment system is installed next. Batteries are housed in battery boxes that are securely attached to sturdy metal battery racks welded or bolted to the car. This ensures that batteries don't take to flight when a driver steps on the brakes or become lethal missiles in an accident.

Fig. 5.7: *The adaptor, shown here, connects the electric motor to the transmission and must be carefully designed and manufactured.*

EV car conversions require deep-cycle lead-acid golf-cart batteries. Once the battery rack and battery box are in place, the remainder of the components are installed. This includes the controller, potbox, main contactor, circuit breakers, battery charger, gauges, and so on. You'll then need to wire the components, including the batteries. Be sure to include safety disconnects, which terminate the flow of electricity in case of an accident.

Bear in mind, too, that when you remove the gas engine, you remove the source of the vacuum (the manifold) that is used to control the power brakes. So, at this point, you may also need to work on the brake system of vehicles with power steering. You can install a vacuum pump and reservoir to operate the brakes. This will cost $200–$400.

The same goes for power steering. You can either adapt the car or truck for manual steering or install a separate vacuum pump for the power steering, which will cost around $800. You will also probably have to beef up the suspension to accommodate the extra weight of the batteries. Most people install a bolt-in set of heavier-duty springs.

You may need to modify the tires. Only radial tires are appropriate for EVs. Low-rolling resistance tires are a good idea, too, as they will let you travel more miles on a single charge. Consider the Michelin or Goodyear high-mileage tires. Once the car is up and running and properly tested, you will need to register the car as an electric conversion. This is usually not a problem.

If the thought of a car or truck conversion seems a bit much for you, you may want to start with a simpler project, like converting a small tractor (Figure 5.8) or a motorcycle (Figure 5.9). Or you can always hire a professional to convert your car for you — provided it is feasible. Companies like Pacific Coast Motors in California (pcmotorsdirect.com), Electric Blue in

Fig. 5.8: *This electric tractor is used to cultivate crops on a very small farm.*

Fig. 5.9: *The front end of this vehicle comes from an ATV; the rear end formerly belonged to a Honda 750. The electric motor came from a forklift.*

Flagstaff, AZ (electricbluemotors.com), and a host of other companies scattered throughout the country will convert your car to electricity. Several companies like LionEV and eeVee Motors specialize in certain car types. LionEV, for instance, converts Ford Rangers whereas eeVee Motors converts Honda Civics.

Is an EV Right for You?

While an EV may sound like the perfect car, be sure to consider this option carefully. Learn as much as you can about EVs, but, most importantly, be sure to realistically compare your needs and the typical driving conditions you encounter to the performance of various EVs. Bear in mind that there are some situations in which EVs don't work well — or will not perform optimally. If you live in an extremely cold environment like northern Alaska, for example, an EV is a poor choice. Cold temperatures reduce the chemical activity of batteries and significantly reduce the range of an EV. In addition, although EVs can work well in wintry climates, use of the heater (an electric resistance heater) can reduce range by 10–50%. Plan for a reduction in range in the winter, and be sure the car can travel the distance required of it.

Very mountainous terrain is also tough on EVs. EVs expend huge amounts of energy climbing steep inclines. In addition, if you are planning on hauling a lot of people or heavy equipment, you probably shouldn't buy an EV. Increased weight dramatically reduces performance. Snowy and gravelly roads reduce mileage, too. Stop-and-go traffic is tough on EVs because it drains batteries.

As a rule, typical electric conversions consume about 0.4 kWh per mile under good road and weather conditions. Adverse conditions will increase that number, possibly to as high as 0.7 or 0.8 kWh per mile. (These numbers are influenced by the weight of the vehicle, driving conditions, and the speed at which the vehicle is driven.)

EVs are ideal for warmer climates with flat terrain. In such areas, you'll get the most miles per charge.

Be sure to consider the number of miles you travel in a typical day. Don't guess. Log your miles every day for a month or two to see how far you travel. Select a vehicle with a range that corresponds to your driving needs. A typical range for EVs is 40–60 miles under favorable conditions, although some EVs can travel 200 miles on a single charge, under good conditions (flat terrain). The Toyota RAV4 EV reportedly can travel about 125 miles on a single charge, under ideal conditions. If you exceed the range of your EV, you'll have to recharge your vehicle at your destination — which means you'll be staying for a while. Or you can take public transportation, rent a car or truck, ask a friend for a ride or to borrow his or her car.

If your travel needs are very modest, for example if you travel 5–10 miles a day to work or a local shopping center, and you live in a warm climate and can travel safely on side streets, you may want to consider a small neighborhood electric vehicle (NEV) like the GEM (Figure 5.10).

When considering this option, you must also consider how fast you must drive. You can't take an NEV out on a highway, as they top out at about 25 mph. Most other EVs can travel at respectable speeds — 65–75 mph — so you won't have to worry about being left in the dust of faster-moving vehicles. A lightweight EV conversion, say, of a Porsche, can travel at 90 mph or higher.

Fig. 5.10: *This vehicle is designed for low-speed use in slow-car-friendly environments — not the open road!*

For the most part, you won't need to worry about acceleration. Just like hybrids and PHEVs, most electric cars can accelerate sufficiently, allowing you to merge into freeway traffic without endangering your life or the life of other drivers. For a detailed look at this subject, I recommend Shari Prange's article "Can an EV Do the Job for Me?" published in issue 91 of my favorite magazine, *Home Power*.

The Pros and Cons of EVs

Electric vehicles offer many benefits over conventional cars. They are cheaper to operate than ICE vehicles. An EV, for example, that gets 0.4 kWh per mile will cost you $1.20 for 30 miles if electricity is selling for 10 cents per kilowatt-hour. Compare that to a conventional car that gets 30 mpg. If gas is priced at $2.50 per gallon, it will cost $2.50 to cover 30 miles. And don't forget that you won't have oil changes or any of the other maintenance costs that go with an ICE vehicle. EVs contain fewer moving parts than ICEs, require less maintenance, and run on fewer lubricants. There's no crankcase oil or engine coolant to worry about.

EVs are extremely quiet, produce no tailpipe emissions, and can be charged at home or at a suitable charging station. Even when powered by coal-fired power plants, they produce fewer pollutants than standard gasoline- or diesel-powered vehicles. When supplied by electricity generated from renewable energy sources, EVs could dramatically reduce air pollution and help us mount a serious attack against global warming.

Since 90% of all Americans travel under 60 miles a day in their cars, EVs could become the commuter vehicle of choice. They're also ideal for running errands. What is more, EVs can be charged at home, making trips to the gas station a thing of the past. Think of the time you could save! It only takes 10–30 seconds to plug an EV into an electrical outlet.

On the downside, the range of most EVs currently available is fairly limited. While there are some EVs that can travel 100 to 200 miles on a single charge, most have a much more limited range — under 60 miles. Some EVs on the market today are fairly slow, too, traveling at 30–40 mph, making them ideal for around-town travel, but inappropriate for higher-speed highway driving. The newest entrants into the market have thankfully solved this problem.

Bear in mind that charging does take time. It requires about 8 to 10 hours to fully charge an EV via a 120-volt outlet, and about half that time when charging from a 240-volt circuit. Some EVs can be charged from either a 120- or a 240-volt electrical outlet, but most are designed for charging from standard 120-volt household current. Newer battery technologies like lithium ion batteries will help reduce charge time and increase range.

Despite these downsides, EVs are here to stay. Thanks to auto manufacturers, we have many affordable choices these days. Expect to see a lot more EVs on the road as people catch on to the marvels of this clean, green form of transportation.

CHAPTER 6

THE PROMISE AND
PERILS OF ETHANOL

Now that we've examined green vehicles, let's take a look at green transportation fuels. Of all the clean, green transportation fuels, ethanol is by far the most widely used alternative to gasoline. If you live in North America, chances are you are burning ethanol in your car right now. That's because many filling stations in the US and Canada sell a blend of 10% ethanol and 90% gasoline, at least during the winter, to help reduce carbon monoxide pollution. If you live in a Midwestern city, chances are there's a local gas station or two selling an 85% blend of alcohol and gasoline, commonly referred to as E85. If you want to make a switch to a renewable biofuel, you may be able to do it right now, provided your vehicle can run on this mix of ethanol and gasoline.

Unlike biodiesel and straight vegetable oil, which are produced on an extremely small scale, ethanol has become big business. In 2007, the US and Brazil, the undisputed world leaders when it comes to ethanol fuel production, generated 6.5 and 5 billion liquid gallons, respectively. Brazil's ethanol industry has created about 1 million jobs. Brazil has attained energy self-sufficiency thanks in part to domestic ethanol production from sugars extracted from the nation's vast fields of

Ethanol Stations

As of January 2009, there were around 1,900 gasoline stations in the US that sold E85. Most of these stations are located in the corn belt states, led by Minnesota with 378 stations. Minnesota is followed by Illinois with 223, Wisconsin with 133 and Missouri with 116. Some states like Alaska, Maine, New Hampshire, Rhode Island and Vermont have no E85 stations at all. Why don't more gas stations sell E85? Unfortunately, E85 must be stored in a separate tank. Installation of a new tanks costs about $60,000.

sugar cane, but also to domestic oil production. Brazil devotes nearly 9 million acres of cropland — about 1% of its total arable land — to sugar cane for ethanol production. In theUS, nearly 25 million acres, or 3.7% of the total arable land, are devoted to corn production to generate ethanol.

Ethanol is not a newcomer among the renewable transportation fuels. In fact, ethanol use as a transportation fuel dates back to Henry Ford. In 1896, Ford designed his first car, the Quadricycle, to run on 100% ethanol. In 1908, Ford's Model T was released. It was capable of running on gasoline, ethanol, or a combination of the two. Henry Ford proclaimed that "ethyl alcohol is the fuel of the future," and he continued to advocate for ethanol as a transportation fuel during Prohibition.

Like other transportation fuels, ethanol has its pluses and minuses. In fact, ethanol has become highly controversial. While farmers and government officials support its production, some scientists and prominent environmental groups have cast doubt on the highly touted benefits of this fuel. Before I cover the objections to ethanol, I'll explore what ethanol is, how it is made, and which crop plants yield this green fuel. Later in this chapter, I'll discuss the flexible-fuel vehicles that burn ethanol and conclude with a summary of the pros and cons of ethanol.

What is Ethanol?

Ethanol is a two-carbon compound with the chemical formula C_2H_5OH (Figure 6.1). The OH at the end of the molecule is an alcohol group. It consists of an oxygen atom chemically bonded to a hydrogen atom. As shown in Figure 6.1, the OH group is chemically bonded to one of ethanol's two carbon atoms.

Ethanol is also known as ethyl alcohol or grain alcohol. Alcohols are water-soluble organic compounds. They evaporate fairly readily and burn cleanly, in part because they contain no impurities like the sulfur contained in gasoline and diesel fuel.

Most readers are familiar with ethanol as the "active ingredient" in beer, wine, and hard liquor. This small molecule has a huge impact on the human body. It alters nerve function in the brain, giving us that feeling of high while reducing coordination

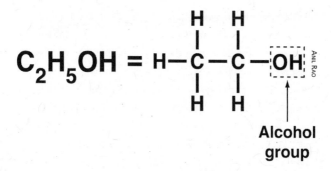

Fig. 6.1: *Ethanol is a two-carbon alcohol that burns inside internal combustion engines.*

and reaction time. It also damages liver cells and is responsible for tens of thousands of highway accidents and about half of all traffic fatalities each year. In sufficient quantities, alcohol can kill a human being outright by suppressing neural activity, especially that of nerves that control breathing and heart beat.

On the plus side, ethanol can be burned in vehicles in place of gasoline, but only in vehicles with properly modified engines (higher compression engines with special gaskets and hoses). More commonly, alcohol is mixed with gasoline at concentrations ranging from 5–95%. As noted above, in the US, many gas stations sell a mix of 10% ethanol and 90% gasoline either throughout the year or just during the winter months. This mix helps reduce carbon monoxide pollution from cars. (Ethanol oxygenates the fuel and reduces carbon monoxide emissions.) No engine modifications are required to burn ethanol at this concentration. As of January 2009, approximately

E85 Fueling Station Statistics

You can tell a lot about a country's commitment to ethanol by the number of ethanol fueling stations per million people in the population. In Brazil, there are 176 stations for every million people. In Sweden, there are 130. In the US, however, there are only 6 stations for every million inhabitants.

You can also tell a lot by the percentage of registered flex-fuel vehicles and the amount of gasoline ethanol "displaces." In Brazil, 12% of all registered vehicles are flex-fuel. In Sweden, it's 2.9%, and in the US it's 2.8%. Unfortunately, most people who drive flex-fuel vehicles in the US don't fill their tanks with E85. Many are not even aware they own a flex-fuel vehicle. As for market share, ethanol's share of the gasoline market is about 50% in Brazil, but only 4% in the US.

1,900 filling stations in the US offered gasoline containing 85% ethanol (E85). Many of these are located in the Midwest, close to the source.

E85 can be burned only in vehicles with specially modified engines, known as flexible-fuel vehicles. Flexible-fuel vehicles burn two or more fuels and are also commonly referred to as flex-fuel or, occasionally, as dual-fuel vehicles. Those that operate on E85 are sometimes referred to as E85 vehicles. E85 vehicles can operate on mixtures of gasoline and ethanol, ranging from 10–85% alcohol or 100% gasoline, as noted in the sidebar.

In cold regions, the amount of alcohol in the fuel is often reduced to 50–70% during winter months to avoid cold-start problems. Reducing the ethanol content also reduces the emission of pollutants during cold starts (discussed below).

In Brazil, all gasoline contains 20–25% ethanol. Over 33,000 filling stations, however, offer a 95% alcohol-gasoline blend (known as E95) for use in flex-fuel vehicles. Some stations also sell 100% ethanol for Brazil's large fleet of ethanol-only vehicles. These cars are equipped with high-compression engines designed to operate on pure alcohol.

Ethanol is also popular in Europe. European nations produce about 90% of the fuel they consume. The largest consumers are Germany, Sweden, France, and Spain. No country is more fond of ethanol fuel than Sweden, which boasts approximately 1,200 filling stations that dispense ethanol.

How is Ethanol Made?

Ethanol is a byproduct of the fermentation of glucose. Glucose is a six-carbon simple sugar with the chemical formula $C_6H_{12}O_6$. It is also known as a monosaccharide. Glucose is present in fruit such as grapes. When fermented, grape juice is turned into delicious wine. The alcohol in the wine is produced from the glucose in the fruit. Glucose is also found in various seeds such as corn, wheat, and barley, which can be used to make beer and hard liquor. Commercially, corn and sugar cane are the two main sources of glucose used to make ethanol fuel. This process also relies on fermentation.

Fermentation is the partial breakdown of glucose by certain species of yeast. This reaction occurs in the absence of oxygen. During fermentation, glucose is broken down to two molecules of ethanol and two molecules of carbon dioxide. For chemically inclined readers, the reaction is this: $C_6H_{12}O_6 \rightarrow 2\ C_2H_5OH + 2\ CO_2$.

Ethanol can then be burned — in the chemical reaction we call combustion. During the combustion reaction, ethanol is broken down into two molecules of carbon dioxide. The reaction also releases heat, which powers vehicles. The released carbon dioxide in this reaction enters the atmosphere, where it can then be taken up by plants grown to produce more glucose used to make ethanol. In essence, then carbon dioxide produced during the combustion of ethanol is continually recycled. It's for this reason that ethanol and other biofuels help reduce carbon dioxide emissions. In contrast, the combustion of fossil fuels releases carbon that has been stored in the Earth's crust for millions of years, producing carbon dioxide. Carbon dioxide released during the combustion of coal, oil, oil byproducts such as gasoline, and natural gas steadily increases in the atmosphere. Rising levels of carbon dioxide, in turn, contribute to global warming.

Glucose used to make ethanol can be extracted from several larger molecules — chemists call them polymers (a polymer is a chemical made up of many smaller molecules). Two of the most common sources of glucose are starch and cellulose, both of which are produced by plants. Starch and cellulose are complex carbohydrates, known as polysaccharides. They consist of many molecules of glucose linked by chemical bonds to form long chains (Figure 6.2). The major structural difference between starch and cellulose is the type of bonds that link the glucose molecules together. In starch, the bond is quite simple to break down. You do it all the time after ingesting starchy foods such as bread, crackers, and pasta. Enzymes in your digestive system break the bonds, releasing glucose that's

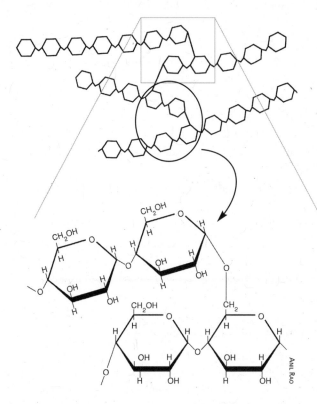

Fig. 6.2: *Many glucose molecules are chemically bonded to form long chains, known as polymers. The most common source of glucose for ethanol production is starch (shown here). Efforts are now underway to tap into the glucose stored in cellulose, a much more common polymer in the plant world.*

absorbed into your bloodstream. Starch can also be broken down by heat in an ethanol refinery.

In cellulose, the bonds joining glucose molecules in the long-chain molecules are much more difficult to break down. Because you lack the enzymes required to break cellulose down, cellulose molecules in foods like celery and carrots are indigestible. We can extract nutrients from such foods, but the glucose molecules in starch aren't available to us. (Cellulose isn't useless, however. It forms the insoluble fiber in our diets that helps reduce colon cancer.) Cows and horses do have the enzymes to break cellulose down to produce glucose, which is why they can subsist on a diet of grasses and grains. As you may know, however, these animals don't produce the enzymes themselves. Rather, they have bacteria in their stomachs that produce the enzymes that can digest cellulose. This results in the release of glucose molecules, which nourishes cows and horses.

Because it is easier to break starch down, very little glucose is currently produced from cellulose. Interestingly however, plants produce a lot more cellulose than glucose. Because cellulose is produced in such abundance, scientists are currently working on ways to extract glucose from this abundant resource as well as the other, smaller plant sugars, notably fructose and sucrose.

Corn vs. Sugar Cane Ethanol

In the US, 97% of all ethanol is produced from corn starch extracted from the kernels. Much of the corn is grown in the Great Plains states like Kansas, Iowa, and Nebraska.

In Brazil, ethanol is produced exclusively from sucrose extracted from sugar cane. Because sugar cane requires a tropical or subtropical climate, very little sugar cane is grown in the US. In fact, only four states have regions suitable for growing this crop: Florida, Louisiana, Texas, and Hawaii. (Louisiana recently opened its first commercial sugar cane ethanol production facility.)

Producing ethanol from sucrose extracted from sugar cane is considerably easier and more efficient than producing it from corn starch. The conversion of sucrose to ethanol primarily requires the addition of yeast, which converts the sugars to ethanol. Making ethanol from corn, in contrast, requires additional cooking and the use of special enzymes. As a result, the energy required to produce ethanol from corn is about twice that required to produce ethanol from sugar cane. The result of this is that sugar cane ethanol production has a much higher net energy efficiency than corn ethanol production — about five times higher.

As noted in Chapter 1, net energy is the amount of energy released from a fuel during combustion minus the energy required to produce it. In the case of biofuels, for example, energy is required to grow and fertilize the crop, transport it to the processing facility, produce ethanol, and ship the fuel to market.

Net energy can be thought of as energy out minus energy in. Fuels are typically compared on a closely related metric net energy *efficiency* basis, often expressed as the ratio of energy in to energy out — which is just another way of looking at the same thing. According to a number of studies, for instance, the net energy efficiency of corn ethanol is 1:1.3 to about 1:1.7. What this means is that every unit of energy invested in the production of corn-based ethanol yields 1.3–1.7 units of usable energy. (Although this doesn't sound impressive, the ratio is actually higher than that of gasoline and diesel fuel.) The net energy efficiency of sugar cane ethanol, depending on the process, is a very impressive 1:8–1:10. In other words, for every unit of energy invested in sugar cane ethanol production, you get 8–10 units of energy out. If it only takes about half as much energy to produce ethanol from sugar cane, why is the net energy efficiency over five times greater?

One reason is that Brazil's sugar cane is a high-yield crop. In fact, sugar cane is one of the most efficient photosynthesizers in the plant kingdom. (It can convert up to 2% of the incident solar energy into biomass. Although that sounds low, the figure is higher than that for most plants, which are around 1% efficient.) As a result, sugar cane yields 700–830 gallons per acre of ethanol, compared to around 400 gallons per acre from corn.

Brazil also enjoys a very long growing season, which further increases output. In addition, Brazilian producers often burn waste material from the cane fields to generate electricity to power ethanol facilities. This lowers the demand for outside energy in ethanol production facilities. (Interestingly, some of these facilities produce excess electricity that is fed into the nation's electrical grid.) Finally, much of the labor is done by hand, too, which boosts the net energy efficiency. (However, there are rumors that slave labor is exploited to produce sugar cane in remote areas of the Amazon basin.) As a side note, other byproducts from the production of sugar cane ethanol are used to fertilize fields prior to planting, further reducing energy requirements.

Because the net energy efficiency of sugar cane ethanol is much higher than that of corn ethanol, Brazilian companies produce ethanol much more cheaply than facilities in the US. The cost to produce a gallon of ethanol from sugar cane is about $0.83 in Brazil; it costs about $1.14 to produce a gallon of ethanol from corn in the US.

Higher net energy efficiency also translates into lower greenhouse gas emissions. It is estimated that the greenhouse gas emission reduction from the use of sugar cane ethanol produced in Brazil is 86–90% when compared to gasoline. The use of corn-based ethanol in the US results in a 10–30% decrease in carbon dioxide emissions. (As you shall soon see, these estimates are pretty controversial.)

What is more, Brazilian officials are looking for ways to increase yields and cut costs even more. For example, new higher-yield strains of sugar cane are being developed. Strains resistant to pests and drought are also in the works. In 2007, Brazilian scientist reported a promising new way to increase ethanol production from yeast. They found that exposing yeast to low-frequency magnetic waves increased the amount of ethanol produced during fermentation by 17%. This treatment also speeds up the chemical reactions, lopping two hours off the standard fermentation times. (Time is money!) Although this discovery took place in the lab, scientists believe that this technique could easily be implemented in large-scale facilities.

Research is also being carried out in the US to increase the output and reduce the cost and emissions of corn-based ethanol. Companies are working on such techniques as cold starch fermentation to reduce energy requirements and boost net energy efficiency — and save money. US companies are also using waste generated during the production of ethanol to produce methane gas that can be burned at the facility to produce process heat and electricity. This could boost the net energy efficiency substantially.

Perhaps an even more important effort is the development of cellulosic ethanol, that is, producing ethanol from cellulose rather than from starch. This technique promises to achieve a much higher net energy efficiency than traditional corn-based ethanol production. (More on this shortly.)

Although ethanol production in Brazil is much less expensive, imports have been limited, in large part because of America's powerful corn lobby — specifically those interested in bolstering domestic ethanol production. To protect their growing and still young industry, lobbyists from the corn-growing states convinced the US Congress to impose a tariff on imported Brazilian ethanol of $0.54 per gallon. Congress also provides the ethanol producers (not the farmers) a $0.51 tax credit for each gallon they produce, which is intended to stimulate domestic production.

Besides the tariff of $0.54 a gallon on imported ethanol, which limits import, Congress also passed legislation that places a 7% maximum limit on imported Brazilian ethanol. In other words, Brazil cannot export to the US more than 7% of the ethanol that was produced in the US in the previous year. Before you chafe at these seemingly unfair policies, say proponents, bear in mind that they are designed to help American farmers and manufacturers who produce our ethanol from an abundant crop: corn. Industry officials, farmers, and government officials are concerned, and rightly so, that an influx of inexpensive ethanol produced from sugar cane and other high-sugar crops might cripple or destroy America's corn ethanol production, which is in an early stage of development.

According to supporters of US policies, Brazil still has huge amounts of land that could be converted to growing sugar cane. Were it to convert this land to sugar cane production, it's conceivable that inexpensive Brazilian ethanol could flood the US market, driving domestic ethanol producers out of business. It's feared that American farmers and manufacturers would be hard-pressed to compete in the ethanol market if restrictions were removed. The Brazilian government has also poured billions, some estimate about $7 billion, into the sugar cane ethanol industry since the mid 1970s — immediately after the first energy crisis. (Brazil took action to achieve energy self-sufficiency and never took its eye off the prize!)

Another reason behind the constraints on Brazilian ethanol imports is that the US is reluctant to trade its long-standing

The Ethanol Corridor

On October 7, 2008, the first "biofuels corridor" was officially opened along I-65, a major interstate highway in the central US. I-65 runs from northern Indiana all the way to southern Alabama. There are now more than 200 fueling stations that carry E85 along I-65. Because of this, it is possible to drive a flex-fuel vehicle from Lake Michigan to the Gulf of Mexico without being farther than a quarter tank's worth of fuel from an E85 gas station.

dependence on foreign oil for a new reliance on foreign renewable fuels. "We are not interested in becoming the Saudi Arabia of ethanol," counters Eduardo Carvalho, director of the National Sugarcane Agro-Industry Union, a producers' group. "It's not our strategy because it doesn't produce results. As a large producer and user, we need to have other big buyers and sellers in the international market if ethanol is to become a commodity, which is our real goal."

Despite the lopsided playing field, Brazil still exports 160 million gallons of ethanol per year to the US.

Cellulosic Ethanol — The Rising Star

As noted earlier, cellulose is produced in abundance by plants. It's the bulk of a tree's trunk, for instance, and makes up a large portion of corn plants and various seed crops. In fact, cellulose is the most common organic compound on the face of the Earth. Tapping into this abundant compound could dramatically expand the amount of glucose that's available for ethanol production in places where sugar cane production is limited by climate. Because cellulose is not digestible, cellulosic ethanol production shouldn't threaten world food supplies either!

Cellulosic ethanol, combined with conventional starch-based ethanol, could enable manufacturers to produce fuel from the entire plant — from the stalks, leaves, and seeds of a corn plant, for instance. This would greatly increase the yields of glucose and dramatically improve the net energy efficiency of production from seed-bearing crops. In fact, it is conceivable that the net energy efficiency of ethanol production from this combined production process could rival, or even exceed, that of sugar cane ethanol.

To preserve food supplies, we could harvest seeds for consumption by humans and livestock, while the cellulosic component would be sent to ethanol plants. This would alleviate concerns over the impact of ethanol production on food supplies and rising food prices, a topic discussed in more detail in the last section of this chapter.

Some experts argue that cellulosic ethanol will simply augment, not replace, corn-based ethanol in the US. However, I believe that the future of ethanol fuel production hinges on the successful development of economical, energy-efficient cellulose ethanol production. Corn should, in my view, be primarily devoted to food production.

In the US, ethanol could be made from cellulose in "stover" — the stalks and leaves of corn plants left over after the corn cobs are removed. Cellulose from wood and straw, which is the leftover stalks of grain crops such as wheat, oats and barley, is another abundant, untapped resource. Several fast-growing and highly productive grasses, switch grass and miscanthus among them, produce enormous amounts of organic material (mostly cellulose) per acre. Hemp, kenaf, and cotton could also be good sources of cellulose. With aggressive and environmentally sensitive development of these and other sources, cellulose could surely augment, or even replace, the production of ethanol from corn. Ethanol production could also be supplemented by production from sorghum, potatoes, sweet potatoes, sugar beets, cassava, sunflowers, fruit, and molasses.

To be sustainable, ethanol production from cellulose will require careful attention to the health and vitality of the soils on which fuel crops are grown. As any agronomist will tell you, soil nutrients and organic matter are vital to the long-term health and productivity of soil. So farmers must take care to return soil nutrients and organic matter to the soil. Without careful replacement strategies, we're simply mining our soils of their natural wealth. Scientists at the National Renewable Energy Laboratory in Golden, Colorado, estimate that two thirds of the biomass of a corn crop could be harvested without damaging the soil. One third would need to be plowed under to ensure an adequate replacement of organic matter. Inorganic nutrients could be replaced by wastes from livestock facilities or perhaps human waste treatment plants, although that can be quite a challenge.

In tropical and subtropical countries, cellulosic ethanol could supplement sugar cane-based ethanol production. The organic materials left over after sugar juices are squeezed from sugar cane stalks, known as bagasse, are currently burned to produce electricity at many refineries, as noted earlier. However, some could be used as a feedstock for cellulosic ethanol production. A dry ton of sugar cane bagasse yields 80 gallons of ethanol, which compares favorably with corn starch, which yields 98 gallons per ton.

Because cellulosic ethanol is in its embryonic stage, it's far behind corn and sugar cane ethanol. In fact, it has only recently left the laboratory and is now being field-tested in a new facility in Louisiana. Stay tuned.

Switching to Ethanol

As noted earlier, you may be burning ethanol in your car, truck, or SUV right now because it's frequently added to the gasoline we buy. Check the gas pump the next time you fill up. There's usually a sticker on the pump that indicates whether the fuel contains ethanol, and how much.

If you would like to burn E85, be sure your car or truck is a flex-fuel vehicle (FFV). One of the main differences between a flex-fuel vehicle and a vehicle that burns gasoline or E10 or E15 is that in the former, hoses are made of materials that can withstand corrosive ethanol. FFVs also come with a special sensor in the fuel line. It analyzes the fuel mixture and controls the fuel injection and timing, automatically adjusting for different fuel ratios (for example, if you fill up on E10 one week, and on E85 the next).

Unfortunately, as noted earlier, many people who drive flex-fuel vehicles are unaware of the fact. In a 2005 survey, for instance, researchers found that 68% of all Americans were not aware that they owned a flex-fuel vehicle, largely because manufacturers haven't done much to distinguish their FFVs from regular vehicles. How do you know if your car or truck is a flex-fuel vehicle?

One way is to check the gas cap. If your car is a flex-fuel car, the gas cap will tell you so. It might even be yellow, indicating that the car can be operated on E85. (Since 2006, many new flex-fuel vehicles in the US are equipped with bright yellow gas caps to signal to their drivers that theirs is an E85 vehicle.) Some gas caps clearly indicate that E85 is *not* to be added to the tank. You can also call the dealership or check the owner's manual for guidance. For a list of flex-fuel vehicles by manufacturer in the US, Europe, and Brazil, visit e85.whipnet.net/flex.cars/.

The First FFV

The first flexible-fuel vehicle produced in the US was the 1996 Ford Taurus. It could run on E85 or methanol — M85.

In 2008, there were over 7.3 million E85 flex-fuel vehicles running on US roads, up from 5 million in 2005. Flex-fuel vehicles are becoming common, especially in the Midwest, where corn is grown and ethanol production is booming. The US government has also been purchasing flex-fuel vehicles for use in its massive fleet of 650,000 vehicles. Currently, nearly 120,000 government vehicles are flex-fuel.

If you car isn't designed to operate on E85, when the time comes to purchase a new vehicle, consider this option. You'll find that most automobiles and light-duty vehicles, such as vans, SUVs, and pickup trucks, are available with a flex-fuel option. Before you purchase a new vehicle, however, be sure that E85 is available in your area. To locate a gas station near you, go to the National Ethanol Vehicle Coalition's website at e85refueling.com/ and click on your state. You may want to check the price, too.

Interestingly, the retail price of E85 varies widely across the US. As you might expect, the lowest prices are found at gas stations in the Midwest — close to the raw materials (corn) and many ethanol refineries. In August 2008, the average spread between the price of E85 and gasoline was about 17%. But national averages are deceptive. In Indiana, the price differential was 35%. That is, ethanol was 35% cheaper than gasoline. In Minnesota and Wisconsin, two pro-renewable energy states, the price differential was 30%. In Maryland, however, the difference was 19%, and in California ethanol was 12–15% cheaper. In Utah, it was just 3% cheaper.

Price difference is important because ethanol contains about 30% less energy per gallon. So when you fill up your car with E85, you can expect to get lower mileage. If E85 is 25–30% less expensive, you're effectively paying the same amount as you would for gasoline. If the price differential is less, you're paying more per mile traveled.

The Pros and Cons of Ethanol

Like any alternative fuel ethanol has its pros and cons. Before you switch to ethanol, you may want to consider these very carefully. Let's start with the pros.

One of ethanol's biggest advantages is that it is a renewable fuel. Moreover, it burns relatively cleanly and could help reduce the emission of the greenhouse gas carbon dioxide, helping reduce the costly social, economic, and environmental impacts of climate change. Ethanol is also produced domestically, which reduces our dependence on foreign oil and helps support farmers and rural economies. In addition, use of this fuel requires very little modification of an engine. Ethanol is also widely available in some locations and reasonably priced. Ethanol is also biodegradable. In an accident, ethanol fires, unlike gasoline ones, are easily put out with water.

On the downside, ethanol production from corn requires massive amounts of land. Corn requires lots of nutrients and, without sustainable fertilization, successive corn crops leave the soil impoverished. If it is not sustainably grown and harvested, widespread corn production to make ethanol could further deplete the valuable agricultural soils of the world.

Corn production requires massive inputs of fertilizer made from natural gas as well as pesticides made from chemicals extracted from oil. Pesticides can pollute ground water and surface waters and kill birds and other beneficial species. Sugar cane requires fewer inputs and is, reportedly, less damaging. However, habitat loss could be a problem if farmers expand their operations to feed the growing biofuels industry.

Corn ethanol also diverts food crops to fuel production and has been blamed for rising prices in the US and less developed countries like Mexico. Some analysts believe that a good part of the price increase is attributable to rising oil prices as well as a rise in the price of corn caused by rich oil sheiks in the

Middle East. They reportedly drove up the price of ethanol by buying corn commodities at inflated prices in an effort to cripple the corn ethanol industry.

High ethanol blends also create problems with cold starts — starting a car on a cold day. That's because alcohol doesn't achieve enough vapor pressure for the fuel to evaporate and spark ignition during cold weather. To avoid this problem, which occurs at temperatures below 59°F, US and European markets have adopted E85. This is the maximum blend that can be used in flexible-fuel vehicles. In regions where temperatures drop even lower, the ethanol content is reduced to 70%. In places where temperatures drop below 10°F, it is recommended to install engine heaters (block heaters) or switch to regular gasoline during cold weather.

Brazilian flex vehicles are built with a small secondary gasoline reservoir to combat this problem. This tank is located near the engine and provides fuel to start the vehicle in cold weather. This is important for people in Brazil's central and southern regions where temperatures drop below 59°F in the winter. Improvements in engine design launched in 2009 could eliminate the need for secondary gas storage tanks.

As noted earlier, ethanol also decreases the fuel mileage in flex-fuel vehicles because fuel economy is directly proportional to the energy content of the fuel. For E10, the effect is quite small. Studies show a very slight decrease in fuel mileage — about 3%. For E85, however, the decline becomes significant, although the performance varies depending on the vehicle. Based on EPA tests run on all 2006 E85 models, the average fuel economy for E85 vehicles dropped 25.6% compared to unleaded gasoline. If you're paying 26% less for E85, there's no economic tradeoff.

Engines that run on 100% ethanol, which could be used to power automobiles, trucks, tractors, and even airplanes, could help solve this problem. Ethanol-only engines have much higher compression ratios, which increase power output and improve fuel economy compared to lower-compression ratio engines in FFVs. These engines, however, are not suitable for gasoline. When ethanol is widely available, high-compression ethanol-only vehicles could be a practical alternative to FFVs. Such vehicles could achieve the same as or slightly greater fuel economy than gasoline engines.

Some studies suggest that ethanol is not as environmentally beneficial as once thought. When ethanol is burned in a car, for example, nitrogen in the air combines with oxygen in the

MIT Develops Ethanol Injection System

Researchers at MIT have discovered an ingenious way to power ethanol cars and trucks. They have devised a dual-fuel direct injection system. One fuel injector injects pure alcohol into the cylinder, the other injects pure gasoline. The fuel mix is controlled by a computer based on engine performance. The fuel is injected into a turbocharged, high-compression ratio, small displacement engine. Though they have fewer cc's, these engines perform like engines with twice the displacement.

In this design, each fuel is delivered to the engine via a separate fuel line. Rather than blending the fuel, there's a tank for gasoline and a much smaller tank for alcohol. The engine runs on gasoline under low-power cruise conditions. Alcohol is injected into the cylinders with gasoline when required, for example when accelerating rapidly. The researchers found that this system reduces gasoline consumption and carbon dioxide emissions by 30%, producing better fuel economy than a turbo diesel or hybrid. It also solves the cold-weather starting problem.

combustion chamber to produce nitrogen oxides. Nitrogen oxides react with water in the presence of sunlight to form nitric acid, which is deposited on the land and water as acid rain and snow. Nitrogen oxides also enter a series of complex reactions that result in the formation of photochemical smog, a toxic brew of chemicals. That said, let's not lose track of the fact that gasoline- and diesel-powered cars and trucks also produce nitrogen oxides, so the lack of other pollutants (like particulates) probably makes ethanol the better fuel.

As for global warming, studies suggest that plowing up new land to grow crops that are used to make ethanol destroys wildlife habitat and reduces the ability of the Earth's ecosystems to incorporate carbon dioxide. (Carbon dioxide is incorporated by vegetation, such as trees and grasses.) Growing fuel crops could therefore increase global warming risks.

If crops are grown on existing farmland, however, and are generated from abundant cellulose, including various types of organic wastes, ethanol could help combat global warming by reducing atmospheric carbon dioxide levels. Stay tuned for more as we move forward.

CHAPTER 7

STRAIGHT VEGETABLE OIL

If you own a diesel car or truck or are in the market for a diesel vehicle, you may want to consider powering your vehicle with vegetable oil. Yes, the exact same oil you use to whip up a healthy stir-fry or prepare artery-clogging fried chicken can be used to power a diesel vehicle. Vegetable oil burns surprisingly clean in diesel vehicles, and is renewable and much more environmentally sustainable than gasoline or ordinary diesel, aka petro-diesel. To run a vehicle on vegetable oil, however, you'll need to first install a conversion kit. Don't just pour vegetable oil in your tank. And remember this, too: only *diesel*

Terminology

Advocates of veggie oil cars use a number of terms, often interchangeably, to describe this fuel. The term *waste vegetable oil*, or *WVO*, is sometimes used, but only refers to vegetable oil that's been discarded from restaurants. *Straight vegetable oil (SVO)* is the term more commonly used. It refers to both waste vegetable oil from restaurants and refined vegetable oil — that is, oil that's not been used previously.

cars and trucks can be converted to run on vegetable oil. Gasoline-powered vehicles cannot. Period.

Diesel Engines and Diesel Fuel

The diesel engine got its start in the 1870s, thanks to the pioneering work of Rudolf Diesel, a bright young engineering student attending the Polytechnic High School of Germany (the equivalent of an engineering college). After learning how inefficient conventional engines were, Diesel decided to design and build a more efficient one. In 1892, after years of work, Rudolf Diesel obtained a patent for the engine that now bears his name.

Although it was considered a breakthrough in engine technology, Diesel's invention was pretty similar to conventional gasoline engines in many respects. For example, the diesel engine is still a two- or four-stroke internal combustion engine. (See sidebar for a description of four-stroke engines.) Like its predecessor, the diesel engine transforms chemical energy in the molecules of the fuel into mechanical energy. As with engines in conventional vehicles, combustion of the fuel inside the cylinders creates a series of small explosions that drives the pistons. The pistons, in turn, are connected to a crankshaft (Figure 7.1). The up-and-down motion of the pistons creates a rotary motion that turns the wheels propelling the vehicle forward. (You can view an animation of a four-stroke diesel engine at auto.howstuffworks.com/diesel1.htm.)

Although diesel engines are similar to gasoline-powered internal combustion engines, there are some significant differences. One important difference is the way in which the fuel is ignited inside the cylinders. In a gasoline engine, as noted in the sidebar, the fuel is mixed with air and drawn into the cylinder during the intake stroke — when the piston descends. The piston then rises during the compression stroke. The compressed air-gas mixture is ignited by a carefully timed electrical

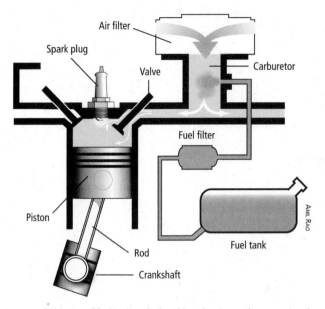

Fig. 7.1: *Combustion of fuel in the cylinders drives the pistons down, turning the crankshaft, creating a rotary motion that turns the wheels of a vehicle.*

spark generated by the spark plugs that project into each cylinder. In a diesel engine, fuel is injected near the top of the compression stroke — as the piston reaches the top of the cylinder. Ignition is not initiated by a spark either. In a diesel engine, it is triggered by heat generated by the compression of air in the cylinder during the compression stroke. (Air heats up when it is compressed.) The explosion caused by ignition of the fuel forces the cylinder down.

Another very important difference between diesel engines and gasoline engines is the way fuel is mixed and injected into the cylinders. In older gas-powered cars, vaporized fuel is mixed with air in the carburetor located on top of the engine. The gas-air mixture is then delivered to the cylinders during the intake stroke. In newer vehicles equipped with fuel injectors, air and fuel are mixed *outside* the cylinders and are then injected into each cylinder. The air-fuel mix is injected into the cylinder during the intake stroke.

In diesel engines, fuel is injected as a fine mist either directly into the cylinders or into precombustion chambers, then into the cylinders. The fuel is not mixed with air prior to injection. Those diesel engines in which fuel is injected directly into the cylinders are known as *direct injection (DI) engines*. The injector sprays the fuel into the cylinders as a fine mist so that it ignites efficiently.

How Four-Stroke Gasoline Engines Work

Four-stroke gasoline engines are so named because they involve four separate phases, called strokes: intake, compression, combustion, and exhaust. During intake, the piston moves down, drawing air into the cylinder through an opening known as the intake port. The opening and closing of the intake port is controlled by the intake valve. In a gasoline engine, air containing vaporized fuel is injected into the cylinder during the intake stroke. (It contains only the equivalent of a drop of fuel each time.) During the next stroke, that is, during compression, the piston moves up, compressing the air-fuel mixture. A carefully timed spark is introduced when the fuel is compressed, creating a small explosion. The force created by the explosion forces the piston down. This is the combustion stroke. It produces the power that turns the wheels. After the combustion stroke, the piston rises again, forcing exhaust gas out via the exhaust port. It is controlled by the exhaust valve. This is the exhaust phase. (To see an animation of this process, go to auto.howstuffworks.com/engine1.htm.)

In an effort to ensure that fuel is evenly distributed in the cylinder, some diesel engines are equipped with precombustion chambers (also called swirl chambers), as mentioned above. Fuel is introduced into the precombustion chamber, and then is injected into cylinders as a fine mist. This helps facilitate combustion and boost efficiency. Engines equipped with such devices are known as *indirect injection (IDI) diesel engines.*

To facilitate starting in cold weather, some diesel engines are equipped with glow plugs. Glow plugs are wires located in each cylinder. They carry electrical current from the batteries each time a diesel engine is started. The flow of electricity through the wires creates heat that warms the inside of the cylinders — in the same way that electricity flowing through the elements of an electric stove or light bulb creates heat. The additional heat aids the ignition by compression when a diesel vehicle cold starts. Preheating the combustion chamber does require some time, usually 10–20 seconds, during which time the driver must sit idly by. Without glow plugs, compression may not increase the temperature of the compressed air sufficiently during cold weather.

Smaller engines like those of diesel lawn tractors and many older diesel car engines typically come with glow plugs to aid in cold-weather starting. However, newer vehicles and larger engines contain advanced computer controls that eliminate the need for glow plugs. These vehicles have sensors that measure ambient air temperature. If it is cold, a signal is sent to the onboard computer. It retards the timing of the engine (when the fuel is injected into the cylinders). As a result, the injector sprays the fuel into the cylinder at a slightly later time during the compression cycle. Because the air is more compressed later in the cycle, it is hotter. This, in turn, allows cold starting without glow plugs.

Diesel cars, trucks, and tractors have historically been powered by diesel fuel, a mixture of hydrocarbons extracted from crude oil (petroleum). Crude oil was formed deep beneath the Earth's surface from ancient marine algae.

Diesel is an oily fuel, much heavier than gasoline. The carbon compounds in gasoline have an average of about 9 carbons each, while those in diesel fuel have 14. Because the molecules are longer, there are more chemical bonds to break. Chemical bonds store energy. So the more bonds, the more energy an engine can extract from a gallon of fuel. In fact, one gallon of gasoline contains 125,000 BTUs (132 x 106 joules). A gallon of diesel fuel contains approximately 147,000 BTUs (155 x 106 joules) — that's about 19% more energy per gallon. The higher energy density of diesel fuel and its higher engine efficiency are the two main reasons why diesel cars and trucks get better mileage than their gasoline-engine brethren.

Converting a Diesel Vehicle to Run on SVO

Vegetable oil can be burned in a diesel engine, but as pointed out earlier, to do so, you — or a qualified diesel mechanic — must first install a conversion kit. If you just fill your tank with vegetable oil and then head out for work, you'll be sorry.

Conversation kits come in two varieties: single-tank and two-tank. As you shall soon see, they contain many of the same components. A high-quality conversion kit will run you about $1,300–$2,000. If you hire a diesel mechanic to perform the work, add another $600–$800.

Both types of kits contain components that heat the vegetable oil. Heating the fuel is necessary because veggie oil is about 10–17 times thicker (more viscous) than diesel fuel. Raising its temperature reduces its viscosity (thickness) so it can flow effortlessly from the fuel tank through the fuel line and the fuel filter and then into the fuel injectors, even on cold days. Heating veggie oil reduces the work the fuel pump must do to move the oil from the fuel tank to the engine. If the oil's

Diesel Engine Fuel Options

Diesel engines can run on three types of fuel: petro-diesel, biodiesel, and straight vegetable oil (SVO). A conversion kit is required to operate on SVO, but not on biodiesel.

too thick, the pump may not be able to deliver a sufficient amount of fuel to the engine, starving it of fuel. Even if the pump is strong enough to move the oil, higher pressure or a vacuum in the system could cause problems elsewhere. It could, for instance, cause fuel lines to rupture or fittings (connections) to leak.

If thick oil makes it to the injectors, its high viscosity will result in a spray that evaporates slowly and unevenly. "The uneven spray leads to incomplete combustion and liquid oil hitting the cylinder walls," notes Forest Gregg, author of *SVO: Powering Your Vehicle with Straight Vegetable Oil.* This causes carbon deposits on various components, which results in numerous problems that reduce engine efficiency, create more wear and tear, and shorten engine life. (I'll explore this topic shortly.)

Single- and two-tank diesel conversion kits contain other components as well, including automatic or manual controls, valves, and meters, which will be described in more detail below.

Two-Tank Conversion Kits

Two-tank conversion kits are widely used in North America and Europe. As their name implies, they require two tanks, the stock tank, which is filled with diesel, and an additional tank for the vegetable oil. The second tank is usually installed in the trunk of a car or the bed of a pickup truck.

Two-tank kits come with a fuel tank heater, fuel line heater, and fuel filter heater. Collectively, they raise and maintain the temperature of the vegetable oil over 160°F. The fuel tank heater is typically a coiled aluminum or copper pipe. (Copper should be avoided as it is a catalyst that causes vegetable oil to react chemically, producing unfavorable byproducts.)

According to the leading manufacturer of conversion kits, Frybid, most kits include only a small heat exchanger in the veggie oil tank to heat the fuel to injection temperature (160°F). Their system utilizes *four* heat exchangers to "assure that all fuel being drawn from the tank is liquefied," according to the company's website. This "lessens the strain on the injection pump and/or fuel pump."

To convert a car to run on vegetable oil with a two-tank system, you must first install the additional tank for the vegetable oil. You must then tap into the vehicle's liquid cooling system to heat the diesel in the tank, the fuel line, and the fuel filter. This is done by installing a tee fitting on the hose that runs from the engine to the cabin heater core. (The heater core is the heat exchanger from which heat is withdrawn on cold winter days when a car's heater is on. It is located just behind the engine compartment, behind the firewall.)

Valves and switches must also be installed. For example, a solenoid valve needs to be installed to allow the driver to change between fuel tanks. The kit should also contain a flushing valve that will clear the vegetable oil from the fuel line before the engine is turned off. Both valves are controlled by dash-mounted switches.

The valves and switches allow the driver to begin and end trips on petro-diesel. Ending a trip on petro-diesel ensures that all the vegetable oil has been purged from the fuel lines, pumps, and fuel filter. This prevents the system from getting clogged by vegetable oil that solidifies on cold days.

A two-tank kit contains a fuel gauge for the vegetable oil tank. It also comes with hoses and connectors. They are required to circulate hot engine coolant to the fuel lines, the fuel filter and the fuel tank, keeping the fuel hot from the tank to the engine.

When a car fitted with a two-tank system is operating, engine coolant flows from the tee fitting (mentioned above) via a one-inch radiator hose. It forms a hot coolant jacket surrounding the fuel line that carries vegetable oil to the engine. This hose also carries hot engine coolant to a copper coil that surrounds the veggie oil filter (some vehicles have electric fuel filter heaters). The coolant then proceeds to the vegetable oil tank where it flows through the heat exchanger(s). The coolant is then returned to the engine by yet another hose where it is reheated.

Operating a Two-Tank Car on SVO

Now that you understand how the system's set up, let's look at how one drives a veggie oil car with a two-tank kit.

When operating a vehicle on straight vegetable oil, the driver must cold-start the vehicle on either the diesel or biodiesel stored in the stock tank. Once the engine and the vegetable oil are warmed up, however, the driver can switch the vehicle to SVO. This usually occurs 5–15 minutes after starting the engine. The car then runs on straight SVO.

Before the engine is shut down, however, the driver must switch from veggie oil back to diesel or biodiesel. The vegetable oil in the fuel lines is then purged by diesel fuel from the stock tank. The unused vegetable oil purged from the lines flows back into the veggie oil tank.

Purging of the fuel lines should occur as a driver approaches his or her final destination. This process ensures that the vehicle is ready for its next cold start — that is, that there's diesel in the fuel line when the car is started again, not vegetable oil. As noted previously, vegetable oil can solidify in cold weather. Even if it doesn't, cold vegetable oil injected into the cylinders does not burn as efficiently if its viscosity is too high. This leads to coking (carbon deposits) and other problems.

Switching can be controlled manually or by an onboard computer, like the Co-Pilot Computer Controller (Figure 7.2). This device takes the guesswork out of running on vegetable oil. The Co-Pilot not only simplifies operation, it keeps drivers from making the most common switching mistakes: converting to vegetable oil operation too quickly and forgetting to switch back to diesel near the end of each trip.

The Co-Pilot's LCD screen displays operation temperatures and notifies the driver with an audible signal when the proper engine temperature is reached so the driver knows when to switch to SVO. In the auto mode, the Co-Pilot automatically switches the vehicle to vegetable oil at the right time.

A timed purge function in this and similar controls activates an automatic backflush that turns on when the engine is switched off. This eliminates the need to switch back to diesel or biodiesel five minutes before shutdown. It also eliminates

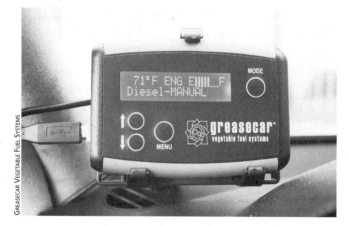

Fig. 7.2: *This device automatically purges the fuel lines of vegetable oil, ensuring optimum performance.*

the possibility of over purging, which results in diesel or biodiesel contamination of the veggie oil.

Two-tank kits are better for longer-distance driving than for short stop-and-start trips. If you use your car principally to run to the grocery store ten minutes from your home, a two-tank kit will provide very little, if any, advantage. You'll be running on petro-diesel most of the time. Running your car on biodiesel or installing a single-tank system would be a better choice.

Single-Tank Conversion Kits

Single-tank conversion kits are the simplest kits on the market, though less commonly used in North America. Single-tank kits employ the vehicle's stock fuel tank to store vegetable oil. No other tank is required. Vehicles run on 100% straight vegetable oil 100% of the time, except perhaps in cold months. A car equipped with a single-tank system therefore starts and stops on SVO — just as if it were diesel or biodiesel.

Single-tank kits were developed in Germany and are extremely popular in Europe. Reportedly, cars equipped with single-tank systems can operate on 100% straight vegetable oil at ambient temperatures as low as 14°F. When temperatures drop below 14°F, the vehicles can be run on winterized diesel or biodiesel. Winterized diesel is diesel fuel with an additive that prevents gelling in cold weather. Supplementary fuel heating is also available for operation on SVO in really cold areas.

Single-tank systems are made by three German companies: Elsbett, VWP and Wolf Pflanzenöltechnik. Table 7.1 lists typical Elsbett conversion kit parts.

The folks at journeytoforever.org, who are unabashed proponents of single-tank systems, claim that these kits are suitable for both direct and indirect injection diesel engines. Other sources suggest that they're not well suited for direct injection engines. When you're starting a DI engine, they argue, the engine and fuel may be too cool to achieve complete combustion, which can lead to serious problems (discussed shortly.) Direct injection engines equipped with electric block heaters may be an exception. Block heaters keep engines warm when a car is switched off in cold weather, so the engine will be warm enough for starting to achieve efficient combustion.

So, if you are considering a single-tank kit, do your research. Talk to people who have experience with conversions. Remember, the main goal in all SVO systems is to reduce inefficient combustion, especially when a vehicle is first started. If the vegetable oil and engine are not sufficiently heated at this time, the result could be a poor spray pattern when the fuel is injected into the cylinder of a direct injection engine. A poor spray pattern results in inefficient combustion that leads to the formation of carbon deposits in the injectors. Carbon deposits on the injectors can also cause the injector needles to stick. (They control the amount of fuel entering the cylinders.) Sticking, in turn, can result in too much or too little fuel being injected into the engine — depending on where the injector is stuck. If too much fuel is delivered, the car produces more pollution. If too little fuel is injected, the vehicle will be underpowered.

Carbon deposits also form on intake and exhaust valves and the exhaust valve stem. Carbon deposits gum up the works, make the engine work harder, and reduce efficiency. Because the fuel burns less efficiently, the car produces more air pollution.

Yet another problem caused by incomplete combustion is that unburned vegetable oil can seep past the rings on the cylinders and drip into the crankcase oil during start-up. This contaminates the engine oil. The rings encircle the pistons and ensure a tight fit in the cylinder so fuel does not leak into the crankcase oil and vice versa. Upon starting, there's a small gap between the piston and the cylinder wall when the engine is

Table 7.1
Parts in a Typical One-Tank Elsbett Conversion Kit

Injector components	Temperature switch
Glow plugs	Cut-off valve
Additional fuel filter	Relays and sockets
Additional fuel pump (sometimes)	Fuel and water pipes
	Fuel hand pump
Coolant/water heat exchanger (to prewarm fuel)	Cabling
	Fitting instructions
Electric fuel heater/filter	User instructions

cold. This gap may allow vegetable oil to leak into the engine oil. As the engine warms up, the pistons expand, closing this gap. Vegetable oil in the crankcase oil is a problem because it reduces the efficiency of the engine and increases wear and tear on engine parts, reducing the life of a vehicle.

Practical Considerations

SVO conversion kits can be installed by an individual with minimal mechanical abilities with tools commonly used by mechanics. If you are not at all mechanically inclined, however, you may want to hire a professional mechanic; better yet, one who has performed SVO conversions before. Call around to local diesel mechanics for references.

While nearly any diesel engine can be converted to operate on vegetable oil, there are some exceptions. It's a good idea, for instance, to avoid vehicles with rubber seals because vegetable oil dissolves rubber (rubber and veggie oil are chemically similar and, in chemistry, like dissolves like). Many older diesels were equipped with rubber seals.

Certain types of diesel engines are also better suited to run on pure vegetable oil than others. Mechanical injection is better than computerized injection, and indirect injection is generally better than direct injection, except perhaps for VW TDIs. Unfortunately, all newer diesel engines come with direct injection engines.

Be sure to look into the type of injector pump, too. Some are better than others when it comes to SVO operation.

To ensure the engine lasts, you'll need to change the oil more often than you're probably used to. Bear in mind, too, that certain vegetable oils work better than others. Mustard and canola oil are among the best because they are high lubricity (low viscosity) oils. In colder climates, use canola, soy, sunflower or corn oil because they don't solidify as quickly as other types of oil.

For reviews of both single- and dual-tank vegetable oil conversion kits, visit FUSEL.com. The site also contains information on ways to obtain designs so you can fabricate your own conversion components from common off-the-shelf hardware, plumbing, and auto parts. Remember, though, that most experts recommend buying a kit rather than going this route.

Securing a Supply of Vegetable Oil

Before you decide to convert your diesel to SVO, be sure you have a reliable supply of vegetable oil. Some individuals purchase refined vegetable oil (unused vegetable oil) in 5-gallon containers or, more commonly, 55-gallon drums. To cut costs, you may want to join with other greasers and buy veggie oil in bulk. Unused vegetable oil costs about the same as petro-diesel when purchased in bulk.

A much cheaper option is to secure a source of *used* vegetable oil. Many restaurants will gladly donate their waste oil and even the containers to carry it away in. Chinese and Japanese restaurants are usually considered the best source because their oils are the cleanest. They are used only once to cook, then discarded.

Oil from other restaurants may be suitable but will likely require more effort on your part to make it usable; mostly, this involves removing food particles. Most restaurants use their deep fryer oil to cook many different foods, bits of which end up in the oil. It's fairly easy to remove the larger bits and pieces by settling or settling and filtration. Flour from breading, however, forms a murky gravy-like suspension in the oil that can be quite difficult to remove. Also, be aware that oil from restaurants that deep-fry meat may contain water, which can seriously damage an engine. Such oil also typically contains animal fat. Animal fats solidify at higher temperatures than vegetable oils and may cause problems in cold weather.

When selecting a restaurant, find one that changes its oil frequently — every week or so. Using the same oil for longer periods changes the chemical composition, making it much less combustible. What you are looking for is oil that is amber in color.

Vegetable oil is delivered to restaurants in five-gallon plastic jugs that are perfect for transporting it to your home. Restaurants may be more than willing to give you some of their used containers. If not, you'll need to find a few containers to transfer oil from the restaurant.

When talking with restaurant owners or managers, be sure to ask about their cleaning operations. Some restaurants clean their fryers with soapy water that is discarded with their waste oil. If their oil is contaminated with soap, you're in for a heck of a time. Kindly request that they place oil and soapy water in separate containers. Take only the oil.

Many restaurants dump their waste vegetable oil in large dumpsters out back. The oils are picked up by rendering companies. They process the waste oil into yellow grease, which they sell. Remember, rendering companies own the bins and the oil in them and have entered into contracts with the restaurants to secure their oil. So don't help yourself — even if the restaurant owner says it's okay. You could be arrested for theft. Instead, talk to the owner or manager to see if they will terminate the contract and give the oil to you instead. (But then you'd better be prepared to take *all* they produce.)

Once you've taken the oil home, let it sit for a week or two undisturbed to allow particles to settle out. You can let your oil settle in 55-gallon steel drums fitted with a standpipe located six inches from the bottom; you'll use the standpipe to drain off the clean oil. Sediment remains on the bottom of the drum. After you drain the clean oil, the bottom sludge can be combined and resettled in another drum. If you obtain high-quality waste oil, it is often not necessary to filter it. Settling works just fine.

Whatever you do, do not underestimate the importance of clean veggie oil. The more energy and effort you put into cleaning your oil, the happier your car will be. Newer vehicles require higher-quality (cleaner) fuel than most older vehicles.

When storing veggie oil, keep it in a cool place, out of the sun in metal or opaque plastic containers that won't allow light in. After a couple of weeks, carefully decant the top 80%. Don't pour it out. Pump it out or drain it using a standpipe. If the oil still contains particles, filter it through cloth bags rated at five microns. The bottom oil from several containers can be combined and resettled. Decant the clean oil and dump what's left in a compost pile where it will decompose naturally. This may attract neighbors' dogs or even coyotes, so be sure to enclose your compost pile.

If you're driving across the country and must stop for veggie oil to fill up your tank, find clean sources and/or clean it as best you can. You may want to keep a spare veggie oil tank on board for settling. (Let it settle overnight while the vehicle is stopped.) Filter the oil, then pour it in your veggie oil tank.

For best results, be sure to check the water and fatty acid content of the oil. If water content is too high, it can seriously damage the engine. Fatty acids can also damage metals in an engine, for example copper and its alloys, such as brass. To learn more about how to measure water and fatty acids, check out Forest Gregg's book, *SVO: Powering Your Vehicle with Straight Vegetable Oil.*

The Pros and Cons of Using Straight Vegetable Oil

Vegetable oil is a homegrown renewable fuel that offers many advantages over conventional petroleum-derived diesel and other biofuels as well. It is non-toxic, non-hazardous, and biodegradable. A spill can be sopped up with kitty litter or sawdust and safely deposited in a compost pile.

Vegetable oil contains more energy per gallon than diesel — about 19% more — and much more than ethanol, natural gas, or propane. Veggie oil also has a respectable net energy yield. Net energy yield is the amount of energy released when a fuel is burned minus the amount it takes to create the fuel (i.e., grow crops, extract and process the oil, deliver it to market, etc.). The net energy efficiency of vegetable oil is twice that of biodiesel and much higher than that of gasoline and diesel. The energetics are even better for waste vegetable oil because you're using a waste product.

SVO burns cleaner than petro-diesel, although there hasn't been much testing on emissions from vehicles using 100% vegetable oil. Emissions are also less damaging to the environment than those released from petro-diesel-powered vehicles. Because SVO contains no sulfur, you won't be spewing sulfur dioxide into the atmosphere as you would with a conventional diesel engine. (Sulfur dioxide reacts with water vapor suspended in the atmosphere to form sulfuric acid that turns into acid rain and snow.) Soot emissions (particulates) are reportedly reduced by 40–60% compared to petro-diesel. Carbon monoxide and hydrocarbons emissions are reduced between 40–60% as well. Carbon dioxide emissions are comparable to other fuels, but are offset by new plant growth. (Plants absorb the carbon dioxide to make new plant material, including oil.) All in all, it's estimated that veggie oil reduces carbon dioxide emissions by over 80%, compared to conventional diesel.

Nitrogen oxide emissions, on the other hand, may be *increased* in vehicles burning straight vegetable oil. (Nitrogen oxides react with water in the atmosphere to produce nitric acid in acid rain and snow.) Fortunately, adjustments to the injection timing and engine operating temperature can reduce emissions significantly — so much so that they can be lower than those from vehicles burning petro-diesel.

On the downside, running a vehicle on SVO today requires one to buy vegetable oil in bulk or, more commonly, to gather fuel from local restaurants. Although the fuel is free, it takes time to collect and clean up the fuel, both of which can be messy.

Although there may be an ample supply of local SVO, supplies from restaurants are necessarily limited — dimming prospects of this fuel as a replacement for gasoline. In 2000, for instance, the US produced more than 2.9 billion gallons of waste vegetable oil, primarily from industrial deep fryers, such as potato processing plants, snack food factories, and fast food restaurants. Even if the entire 2.9 billion gallons were collected and used as fuel, it would replace less than 1% of US oil consumption.

That said, refined vegetable oil could also be extracted from crops grown specifically to produce SVO, though that would bring with it some of the same problems presented by ethanol and production of other fuel crops.

Another downside of SVO is that there is no infrastructure for it — as there is for ethanol and, to a lesser extent, biodiesel.

SVO is currently a do-it-yourself venture or, at best, a small-group cooperative endeavor. Because the amount of used waste vegetable oil is limited, it seems unlikely that infrastructure will be created to provide this fuel commercially.

Another downside emerges on long-distance trips. If you're traveling across the country, you won't find gas stations offering SVO. Stopping to gather vegetable oil can be very time consuming. If you can't find a supply, you have to switch to petro-diesel or biodiesel.

There are also some legal issues to consider. In the US, for instance, it is illegal to convert a car to run on straight vegetable oil under US EPA guidelines. Doing so could affect the emissions of a vehicle and is considered illegal tampering. Although the EPA has not fined anyone for converting a vehicle to SVO, it is still illegal. It's also illegal to sell kits unless they've been certified by the EPA. So far, there are only two companies seeking certification, which is an extremely time-consuming and expensive process.

It is also illegal in the US to sell straight vegetable oil as a fuel. Those making their own vegetable oil are required to pay federal and state highway taxes on the fuel they produce and consume. (Although it is illegal to convert a car, they'll gladly accept your tax money.)

Taxes on SVO vehicle fuel vary from one country to the next. Some revenue departments may not even be aware of its use or may believe that it is not significant enough to collect taxes on. In Australia, it is illegal to produce any fuel for sale unless a license is granted by the federal government. Failing to do so can result in a fine of up to $15,000–$20,000 — and can also involve jail time!

As a final note, converting your car to run on vegetable oil may void manufacturer warranties. Contact your dealer or the auto manufacturer to determine if this is the case.

Conclusion

News stories about people traveling across the country on veggie oil are alluring to those of us who are interested in sustainable transportation. It should be clear by now, however, that it is not quite as simple as it looks.

You need to look into your options very carefully to make the right choice. I strongly recommend that you read more, especially Forest Gregg's book.

When purchasing a kit, bear in mind that climate plays a big role in the success of an SVO vehicle. Warmer climates are better than colder climates. In fact, in warm climates, or during warm weather in colder climates, cars can be started and run completely on vegetable oil. Purging the fuel line of SVO may not even be necessary in such instances.

If you convert your vehicle to SVO, you may want to consider using a fuel injector/piston cleaner to remove carbon deposits. Pour one 12-ounce bottle into the fuel tank every six months.

Finally, when shopping, avoid SVO systems containing copper parts, for example copper tank heat exchangers. (That's not because the oil will damage the copper, but rather the copper will catalyze chemical reactions in the oil that result in byproducts you don't want in your tank and engine.) When shopping, choose wisely. Buy the best kit you can afford.

If you're in the market for a diesel car, be sure it's in good operating condition. If you buy a lemon, it will still be a lemon after you convert it to veggie oil, note the folks at Greasecar. For guidelines, see their 23-point inspection checklist posted on their website at greasecar.com. It will help you make the proper selection of a diesel vehicle for conversion.

CHAPTER 8

BIODIESEL

My first encounter with biodiesel occurred in Iowa at the Iowa Renewable Energy Fair in 2001. I'd given the keynote address that day and was taken to dinner by members of the board of directors of the event's sponsor, the Iowa Renewable Energy Association. The gentleman who drove me to dinner, Marc Franke, took me in his VW powered by 100% biodiesel. After dinner, Marc dropped me and a few others off at the fairground. We chatted for a while before he took off, standing near the back of his car while the engine was running. (We'd gone back there to smell the exhaust, which had a faint odor of French fries.) Soon after our sniff test, we found ourselves engrossed in conversation. Time passed by quickly. I realized that we had been talking for 20 minutes, standing close to the exhaust. Had the car been burning conventional diesel, we'd all have succumbed to the fumes. As it was, however, we didn't even notice.

Impressed, I decided to learn more about this renewable fuel. In a class on energy that I taught at Colorado College with Professor of Chemistry Sally Meyer, she taught students how to make biodiesel in the lab.

A few years later, Professor Meyer made a batch for one of my classes — to show my students how it was made. She gave me the output from the college's small reactor. I took the five gallons to my office before taking them home to power my lawn diesel tractor, which I use to plow my driveway.

Unfortunately, the five-gallon container was pretty full, and when placed in my warm office, the fuel expanded. As a result, some of the biodiesel oozed out of the top of the jug and spilled onto my carpet. I asked the secretary to call the janitorial staff. However, we decided to tell them it was vegetable oil rather than biodiesel. Even though biodiesel is perfectly safe, we were afraid that they might hit the panic button, assuming biodiesel was chemically similar to petro-diesel, which is pretty toxic stuff. We imagined they'd want to call in the hazardous waste team to sop up the mess. The cost of the cleanup, including new carpeting, might have also raised the ire of a dean or two. As it turned out, housekeeping showed up, sopped up the "veggie oil", and we were done.

In this chapter, we'll examine biodiesel. You'll learn what it is, how it is made, and how you can make your own. We'll also explore the pros and cons of biodiesel.

What is Biodiesel?

Biodiesel is a clean-burning, non-toxic fuel (Figure 8.1). It can be burned in almost any diesel engine, from cars and trucks to buses and trains. Unlike vegetable oil, there's no need to modify an engine to burn biodiesel. The only exception is older diesel vehicles that are equipped with rubber seals and hoses. (By older I mean manufactured before the late 1970s.) Small amounts of methanol in biodiesel dissolve rubber, causing tiny leaks. Leaks, in turn, create a mess in the engine compartment and a potential fire hazard. Don't despair, however, if you own one of these vehicles. The rubber hoses and seals can be replaced with synthetic counterparts that easily stand up to biodiesel. Chances are an older vehicle has already been refitted

Fig. 8.1: *Biodiesel is available at many gas stations throughout the US and can be burned directly in many diesel vehicles.*

with biodiesel-compatible components, as rubber hoses and seals typically give out.

Biodiesel is a renewable fuel made from vegetable oil — virtually any vegetable oil. It can also be made from animal fat, including human cellulite and belly fat. (Think of the cars we could fuel with biodiesel from human fat!)

Commercially, biodiesel is usually made from refined (unused) soy or canola oil (also known as rapeseed oil). Most homebrew biodiesel, however, is made from waste vegetable oil, that is, previously used vegetable oil typically obtained from local restaurants. That said, some homebrewers make theirs from refined oil purchased in bulk, usually 5-gallon carboys or 55-gallon drums.

As you may recall from your high school or college biology course, fats and oils contain organic molecules known as triglycerides. As shown in Figure 8.2, a triglyceride consists of a three-carbon backbone, glycerol. Attached to it are three long-chain fatty acids.

To make biodiesel, the triglycerides must be broken down chemically into smaller molecules, notably glycerol and fatty acids, as shown in Figure 8.3. When entirely broken down, each triglyceride molecule yields three molecules of fatty acid and one molecule of glycerol. The glycerol is removed and

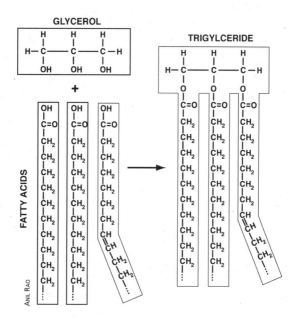

Fig. 8.2: *Triglycerides are lipids made from glycerol and three fatty acid molecules.*

added to soap or discarded (it can be added to a compost pile). The fatty acids are extracted and become biodiesel.

How Is Biodiesel Made?

Biodiesel can be made at any scale — from a 250-ml flask to a 10,000-gallon commercial reactor vessel. No matter how it's made, the steps are the same.

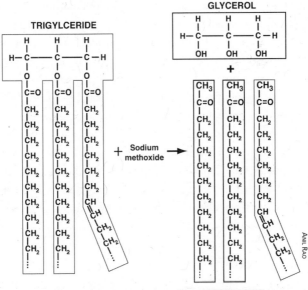

Fig. 8.3: *This simple chemical reaction lops off the fatty acid molecules of the triglycerides, producing the fatty acids of biodiesel that burn nicely in a diesel vehicle. As shown, the fatty acid molecules are cleaved off very close to the glycerol backbone of the triglyceride molecule. During this chemical reaction, a CH_3O group from methanol (CH_3OH) is inserted on the end of fatty acid molecules where they once attached to the glycerol molecule. The product is known as fatty acid methyl ester (FAME). It's called a methyl ester because CH_3 is a methyl group. The chemical bond that forms is called an ester. The chemical reaction is known as a transesterification reaction.*

The production of biodiesel begins with the addition of a well-blended mixture of lye and alcohol to vegetable oil. Lye is an alkaline solution, consisting of either sodium or potassium hydroxide. The alcohol most commonly used to make biodiesel is methanol (aka methyl alcohol or wood alcohol). Other types of alcohol, such as ethanol (grain alcohol) and isopropyl alcohol (rubbing alcohol), can also be used. When sodium hydroxide and methanol are mixed, they form sodium methoxide.

Sodium methoxide reacts with the triglycerides in vegetable oil or fat. During this reaction, the fatty acids are "lopped" off the glycerol backbone of the triglyceride molecule. The products of the reaction are glycerol and free fatty acids.

To make biodiesel, we must first thoroughly mix methanol and sodium hydroxide (a strong base) in a separate container. The sodium methoxide is then added to a container of vegetable oil or fat. Then the mixture of vegetable oil and sodium methoxide is heated to 130°F for one to four hours and stirred to ensure that the reaction proceeds to completion. (The time required varies with the amount of oil, the wattage of the heating element, and the efficacy of the agitation.)

Once the reaction is complete, the agitator is turned off. Over the next 24 hours, biodiesel consisting of the fatty acid methyl esters floats to the top of the reaction vessel. (It's is honey-yellow in color.) Glycerol sinks to the bottom of the reactor, where it can be drained off. (Glycerol has a dark color.) An intermediate layer containing soaps and partially broken-down trigylcerides, known as monoglycerides and diglycerides, often forms. (A monoglyceride consists of one fatty acid molecule attached to a glycerol molecule. A diglyceride consists of glycerol and two fatty acids.)

After the glycerol and the other impurities are drained off, the biodiesel is washed with water, either as a spray or by bubbling it through the biodiesel. It is then allowed to resettle. Any

Winterizing Biodiesel

To prevent gelling, homebrewers can purchase a winterizing agent at auto parts stores to add to their biodiesel in the winter months. This is the same agent used in petroleum-based diesel.

remaining glycerol and impurities sink to the bottom and are removed. The biodiesel is then washed again, after which it is ready to use.

Biodiesel can be burned in a diesel vehicle at full strength in warm weather or in warm climates. One hundred percent biodiesel is known as B100. Like straight vegetable oil, biodiesel will gel when the mercury falls. (It begins to gel at around 45°F.) Therefore, users in regions with temperate climates, like most of the US and Europe, typically blend biodiesel with conventional petro-diesel. (Although conventional petro-diesel also gels, refineries add anti-gelling agents as the temperature declines in the fall and winter.) As the mercury drops, a home-brew diesel is converted to B60 — a 60% biodiesel/40% petro-diesel mix. As the outside temperature decreases even further, B40 — 40% biodiesel/60% petro-diesel — may be required. In really cold weather, the concentration of biodiesel is reduced even further, to B20. B20 is available year-round from hundreds of filling stations in the US and other countries.

Making Your Own Biodiesel

The easiest way to green your ride is to buy biodiesel from local filling stations. You can locate them by visiting biodiesel.org/buyingbiodiesel/retailfuelingsites/ and clicking on your state.

If you're interested in making your own biodiesel, either to use as fuel or just for the fun of it, you may want to start small, with a two-liter soda bottle as your biodiesel reactor. Detailed directions are available online at biodieselcommunity.org/makingasmallbatch/.

Follow the directions, and be very careful when handling methanol, lye, and sodium methoxide. Read safety instructions. Above all, remember that methanol is a poison. It can be absorbed through the skin or enter the body via inhalation. If you ingest or inhale too much, it can cause blindness, or even death. Methanol is also as flammable as gasoline or diesel. Be sure you make biodiesel in a well-ventilated space away from flames or sparks. Sodium hydroxide is caustic and can cause severe burns, or even death if ingested. So *please* be very careful — we need you around to help create a sustainable energy future!

When making biodiesel, wear rubber gloves, a long-sleeved shirt, and safety goggles or glasses. Don an apron for additional protection. Whatever you do, do not breathe in any vapors. And be close to running water so you can wash off any accidental spillage. It's also a good idea to have a fire extinguisher handy, just in case. Keep children and pets away from your operation.

Be sure to rid veggie oil of water before you start. It can be present, for example if the bottle containing vegetable oil has been open for a while. The online directions will explain how to get rid of the water. If you're using a brand-new (unopened) bottle of veggie oil, there shouldn't be any water in it.

Whatever you do, *don't mix methanol and sodium hydroxide in the plastic soda bottle*. It will eat right through it. Mix it in a glass container, like a measuring cup. You can, however, mix the veggie oil and sodium methoxide in the plastic soda bottle.

If you are interested in making a small batch, you can purchase the chemicals through a local scientific supply store or online from scientific chemical supply outlets. Methanol is sold in hobby shops as model airplane fuel. You can also get it at an auto supply store; HEET brand fuel line antifreeze (in the yellow bottle) is methanol.

Lye (sodium hydroxide, or NaOH) is the ingredient in ordinary drain cleaner like Drain-O. Whatever product you use, make sure the container says it contains *only* sodium hydroxide. Not all drain cleaners are 100% lye. Liquid-Plumr, for instance, contains sodium hydroxide, but also sodium hypochlorite (bleach) and other chemicals (sodium silicate and surfactants). You don't want to use any formulation other than one containing 100% sodium hydroxide. Especially avoid drain cleaners that contain NaOH and aluminum pellets.

To make enough biodiesel to power a car, truck, or tractor, you'll need to scale up your operation. You can buy a commercially produced biodiesel reactor like the Fuelpod 2 (Figure 8.4), although it will cost you $3,000–$5,000, or you can purchase a kit that comes with all the parts (valves, fittings, tubing, etc.; all you have to do is supply the electric water heater).

Fig. 8.4: *This self-contained, highly automated biodiesel reactor manufactured by Green Fuels can be used to produce biodiesel in your garage. This unit costs $4,995 and comes equipped with a dry wash system. It uses an air-driven pump for maximum safety.*

GREEN FUELS AMERICA

That's a much cheaper option. Expect to pay $300–$600 for a kit. They're available online. You can also build your own reactor from parts you purchase locally. For advice on building your own reactor, check out the articles and books listed in the Resource Guide at the end of this book.

Reactor Equipment

However you build your reactor vessel, it will need a heat source and must be designed so that it will thoroughly mix the reactants for several hours. Your main objective is to be able to heat the sodium methoxide/veggie oil mix, stir it, and do it safely.

Some commercially available small biodiesel reactors for individual use — like the Fuelmeister — employ plastic tanks for mixing methanol and reacting sodium methoxide with the vegetable oil. While plastic is less expensive than other materials, some biodiesel experts believe that plastic is a bad choice. It may be cheap, light, and translucent (so you can see the contents), but it can catch fire if the biodiesel leaks out and ignites.

To avert this problem, some suggest *not* heating the biodiesel reactor. While this does reduce fire hazards, it also produces a much lower-quality fuel. The long-term effects of burning a poorer-quality fuel in a diesel engine are unknown, but it's too risky to gamble.

Plastic reactors also require more energy to "drive" the chemical reaction because a considerable amount of heat will be radiated from the uninsulated walls of the reactor. Without insulation, the heating time increases by many hours. In addition, uninsulated tanks may be a bit cooler, which can result in a lower-quality fuel.

When you're designing a reactor, it is a good idea to create a closed system — that is, a system that allows no methanol fumes to escape. Methanol fumes, as noted above, are potentially dangerous. Moreover, no respirator or filter will stand up to methanol fumes for more than a few minutes without methanol breakthrough (methanol makes its way through the filter medium).

Many people use electric water heaters to create biodiesel reactors (Figure 8.5). Do not use a gas or propane water heater. You don't want a flame near your biodiesel operation.

Electric water heaters can be obtained for free, but be sure to get one that doesn't leak. (They are often discarded because they *are* leaking.) Also, remember that electric water heaters

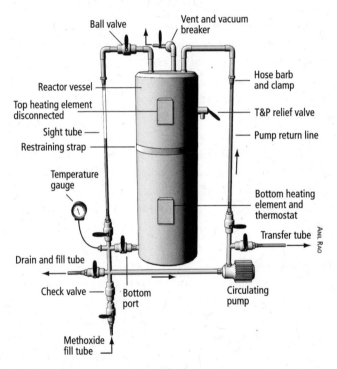

Ball valve

Vent and vacuum breaker

Reactor vessel

Top heating element disconnected

Sight tube

Restraining strap

Temperature gauge

Hose barb and clamp

T&P relief valve

Pump return line

Bottom heating element and thermostat

Transfer tube

ANIL RAO

Drain and fill tube

Check valve

Bottom port

Circulating pump

Methoxide fill tube

Fig. 8.5: *Biodiesel reactors can be made from used electric water heaters. The Appleseed biodiesel reactor, described in* Home Power *magazine (issue 107), is a terrific design that's been improved on over the years. This reactor is made from an old electric water heater. This system can be built for $250–$600.*

require 240-volt AC electricity. You'll need to have 240-AC service and an outlet in the location in which you're going to be making biodiesel. Electric water heaters contain two heating elements, one near the bottom of the tank and the other near the top. Be sure to disconnect the top heater. It will burn out if not submersed in liquid, and you very likely won't fill the entire tank with biodiesel.

If you adapt an electric water heater to make biodiesel, be sure to remove the anode rod. The anode rod is a zinc or magnesium rod suspended inside the tank from the top. It helps lengthen tank life (it reduces rusting of the steel tank). The anode rod is sometimes clearly marked but may be hidden by a plastic plug in the top of the tank.

When designing your own system, remember that safety is the most important goal. Even if your budget constrains your ambitions, don't sacrifice on safety. Methanol is highly reactive and can combust quite easily.

Once you obtain a clean, watertight water heater, you'll need to fit it with pipes and a pump, and install valves and fittings. The pump is used to fill the reactor vessel with vegetable oil. It is also used to agitate the mix when it is being heated and to draw off the glycerol and the biodiesel when the reaction is done. The valves allow the operator to control the flow of reactants and products.

When building your own system, be sure to use biodiesel-compatible pipes, fittings, and valves. For pipe, select non-galvanized mild steel (also known as black pipe), or, if you can afford it, use stainless steel pipes and fittings. Avoid copper, zinc, and iron because they all can catalyze the oxidation of biodiesel and thus shorten its shelf life. Storage is generally not a problem with biodiesel homebrewers, as they use the fuel fairly quickly. Oxidation does become a problem if fuel is stored for long periods, however.

Although it is pretty easy to avoid copper, it is much more difficult to avoid zinc because it's used to galvanize pipe and is also found in brass valve bodies. Cast-iron fittings are also hard to avoid if you're building a reactor on a budget.

Get it Right!

Although biodiesel production is pretty easy, once you get the hang of it, creating a high-quality fuel can be a challenge. If not made properly, the fuel can damage an engine. If too much lye is added, a thin skin of film forms on the top of the fuel. You can skim this film off, but it will form again and again. If this fuel is burned, it can gunk up the engine. Problems also arise if too little lye is used. The reaction won't take place fully and you could end up contaminating your fuel with monoglycerides and diglycerides, which don't burn as well as fatty acid methyl esters.

While biodiesel production seems pretty straightforward, you need to do it carefully. To learn more, I suggest you read some of the books listed in the Resource Guide. To learn about obtaining oil to make biodiesel, be sure to read the section entitled "Securing a Supply of Vegetable Oil" in Chapter 7.

The Pros and Cons of Biodiesel

I'll start with the bad news.

Although biodiesel is a promising green transportation fuel, its production requires the use of toxic chemicals, notably lye and methanol. Another huge problem is that the oils used to make it are currently limited. In fact, government studies indicate that biodiesel could provide a paltry 7% of the diesel fuel consumed in the US each year — and then only if *all* the waste oil generated in this country were fully utilized.

Another downside of biodiesel (specifically biodiesel made from refined vegetable oils) is that it requires land to grow crops, which means it invokes all the same evils of other plant-derived energy sources, like ethanol. In addition, many tropical forests are currently being ravaged to plant palm trees to generate palm oil. It is extracted to satisfy the growing markets for biodiesel, especially in Europe.

On a positive note, biodiesel is a non-toxic, biodegradable, and renewable energy resource that can be made from almost any seed crop. Because biodiesel can be made from refined or waste veggie oil, it helps put a messy waste product to very good use.

Although the supply of used vegetable oil is limited, biodiesel can also be made from algae. In fact, algal production may be one of the most promising energy sources. Some proponents envision large algae ponds located near sewage treatment plants. The algae would dine on organic wastes from human sewage and would be harvested to produce oils. This scheme could produce massive amounts of oil because algae produce much more oil per acre than *any* land-based crop!

Biodiesel can also be produced from livestock wastes (manure), which are abundant in many parts of the world. Chicken and turkey manure, for instance, can be used to make biodiesel. All these sources, when combined, could dramatically increase biodiesel's contribution to global fuel supplies.

Burning biodiesel reduces emissions of virtually all air pollutants and could help us combat global warming, urban air pollution, and acid rain. The only air pollutant that increases with the use of biodiesel is nitrogen oxides. These chemicals combine with water to form nitric acid, which falls as acid rain. They also participate in the complex series of chemical reactions in the atmosphere that result in the formation of photochemical smog.

Although nitrogen oxide emissions could increase with the use of biodiesel, there are ways to thwart the problem. Simply changing the timing in a diesel engine helps to reduce nitrogen dioxide emissions from an engine. In Europe, where diesel vehicles are extremely popular, cars are equipped with catalytic converters that reduce nitrous oxide emission dramatically. Although the use of a catalytic converter may result in a very slight loss (about 5%) of engine power, this reduction may be compensated for by the higher lubricity of biodiesel. Because the engine is more lubricated, it runs more smoothly.

According to an EPA-approved study, burning biodiesel in an unmodified diesel engine reduces the emission of banned carcinogens by up to 90%. It also reduces sulfur emissions by nearly 100% and global warming gases by more than 90%.

Biodiesel is also much safer to store than conventional diesel or gasoline. That's because it has a higher flash point (it ignites at a higher temperature) and is much less volatile.

Biodiesel is also a lubricant that reduces wear and tear on an engine and makes it run more quietly. Homebrew biodiesel is fairly inexpensive and it's easy to make. It can also be purchased at many filling stations, especially in the Midwestern US.

Biodiesel is also non-toxic, making it an excellent fuel source for boats, including yachts and cruise ships. (At least one cruise ship line currently uses biodiesel.) Because it is non-toxic, spills from ships are much less harmful to aquatic life than petro-diesel spills; it also degrades much faster than petro-diesel.

Another benefit of biodiesel is its higher net energy efficiency. As noted in Chapter 7, net energy is the energy produced by a fuel when burned minus the energy it takes to grow and harvest the crop, extract the oils, make the fuel, and transport it to market. Studies suggest that conventional diesel has a net negative energy yield. That is, it takes more energy to make diesel than you get out. Because of the processing required to make biodiesel — the use of methanol and sodium hydroxide (both of which require energy to generate) — the net energy efficiency of biodiesel is about half of that of SVO (which is 6.4). According to studies, the net energy efficiency of biodiesel made from soybean oil is about 3.2. If it is made from canola oil, which produces more oil per acre, the net energy efficiency is 4.3. Even though biodiesel is not as impressive as SVO, it's much better than diesel.

Conclusion

If you're interested in greening your wheels, homemade or commercially produced biodiesel may be the answer. If you own a diesel vehicle and buy commercially manufactured biodiesel, fill your tank with B20 and go. It's that easy. Well, almost.

Biodiesel is a pretty good solvent and will "scrub" the gunk that has settled in your fuel tank and remove the crud coating the internal linings of your car's fuel lines. As a result, you can expect your fuel filter to clog up pretty quickly. Keep an eye on it. To avoid running into problems, notably starving your engine of fuel, change the fuel filter frequently at the beginning. Once the tank and fuel lines are clean, you can return to routine fuel filter changes.

If you're making your own biodiesel, start with B20, then after a while replace the fuel filter, move to B40, and so on. Once the tank and fuel lines have been scrubbed clean, you can run on B100.

Happy motoring!

CHAPTER 9

HYDROGEN AND HYDROGEN FUEL CELLS

Imagine a transportation fuel that could be made from water, the most abundant chemical compound on the planet. Imagine, too, that this fuel, when burned or processed, would regenerate the very same material used to make it: water. Imagine, too, that when this fuel is used to power trucks, cars, and buses, it produces no pollutants whatsoever. The fuel, of course, is hydrogen.

Not only does hydrogen come from an abundant natural resource, it has the highest energy content of any fuel known to humankind (based on weight). It has become one of the shining stars of green transportation fuels. In fact, many clean-car advocates and world leaders are pinning their hopes on the successful development of this seemingly perfect fuel. But is hydrogen all that it's cracked up to be? Will it become a major source of transportation fuel in the future? Or will the laws of physics banish hydrogen to the scrap heap of exciting but unrealistic pipe dreams?

What is Hydrogen?

Hydrogen is the most common element in the universe, but there's one hitch. While the element hydrogen is found in great abundance, hydrogen *gas* (H_2) is quite rare. Sure, it is produced by a number of natural processes and is found in minute quantities in natural gas and even biogas. But because it is highly reactive, hydrogen gas is present in very limited quantity. There are no deposits of it sitting around waiting to be tapped.

Hydrogen gas is non-polluting and as safe as gasoline — maybe safer. It can also be produced just about anywhere, either from hydrogen atoms stripped from hydrocarbons like methane or from water molecules. Because hydrogen gas can be derived from water, which covers two thirds of our planet's surface, it could be the perfect fuel.

As a gas, hydrogen is considered the ultimate "clean energy carrier." (It's more properly referred to as a "carrier," rather than a fuel, because it is not naturally occurring, like coal or natural gas or oil.) It can be burned directly, for example in a furnace or a stove or even a car. Hydrogen gas can also be fed into a device known as a fuel cell, a device that, through a bit of chemical wizardry, produces electricity.

Hydrogen is currently generated in massive quantities for industry. It is used in many disparate industrial processes, such as hydrogenating vegetable oils to produce margarine and making fertilizer. Even the rockets that NASA sends into outer space rely on hydrogen. In addition, hydrogen fuel cells on the space shuttle consume hydrogen to produce the electricity that powers the vehicle's lights, pumps, and instruments. The only emission of the onboard fuel cells is water, which is consumed by the crew.

How Hydrogen is Made into Fuel

Making hydrogen gas to use as a fuel involves stripping hydrogen atoms from various molecules, such as water. Unfortunately, it takes a lot of energy to separate hydrogen atoms from other atoms to which they are tightly bound. Once they *are* pried loose, hydrogen atoms quickly react with one another to form hydrogen gas.

Hydrogen can be extracted from a variety of substances besides water, including fossil fuels and some forms of organic matter. Today, manufacturers generate 40 million tons of hydrogen worldwide. Most of it comes from natural gas (methane, CH_4) via *reforming*. In this process, natural gas is combined with steam — superhot water vapor. The reaction yields hydrogen gas and carbon dioxide.

Methane reforming is the cheapest of the half dozen or so manufacturing processes that generate hydrogen. Some consider it the most likely candidate for generating the hydrogen needed to power cars, buses, and trucks powered by fuel cells. Hydrogen can also be generated from methanol, coal, and other hydrogen-rich fuels. Unfortunately, all these processes produce significant amounts of carbon dioxide. From an environmental standpoint, extracting hydrogen from hydrocarbons is a much less desirable option than extracting it from water.

Electrolysis is the process used to strip hydrogen gas from water molecules. During electrolysis, an electric current is passed through water. This causes water molecules to split into hydrogen and oxygen gases. These molecules can then be fed into a fuel cell to generate electricity to power an electric car equipped with an onboard fuel cell.

Unfortunately, the electricity required to split water into oxygen and hydrogen has to be generated somewhere else. One simple and inexpensive way (if you ignore the environmental impacts) is to generate electricity by burning coal or natural gas. Proponents of nuclear power assert that the intense heat generated at nuclear power plants could be used to split water to make hydrogen. This option, however, is fraught with problems, among them the extremely high construction costs of nuclear power plants, the exorbitant liability in case of an accident, waste disposal, and security.

Recognizing the pitfalls of producing electricity from fossil fuels and nuclear power plants, many proponents of hydrogen have pinned their hopes on generating electricity from renewable energy technologies, such as solar electricity, wind, hydro power, geothermal, and biomass. The use of these renewable energy technologies could help render hydrogen production carbon-free — or at least dramatically reduce its carbon footprint.

Is There Enough Water to Make Hydrogen Fuel?

As noted in previous chapters, the production of biodiesel, straight vegetable oil, and ethanol is hampered — at least now — by a lack of sufficient feedstock. Put another way, the world's farmers don't grow enough fuel crops to generate a sufficient amount of these fuels. And even if they did, the economic repercussions on food prices could be serious. Especially vulnerable are the world's poor, who depend on cheap grain crops to meet a good part of their nutritional needs.

To produce sufficient amounts of the biological feedstocks needed to meet the world's astronomical transportation fuel needs, nations would need to convert vast acreages of cropland and wildlands to the cultivation of fuel crops — a move that could have serious ecological impacts. Even then, we may not be able to produce enough fuel to replace the massive quantities of gasoline, diesel, and jet fuel currently consumed by humankind's millions of motorized vehicles. What about hydrogen? Is there enough water to yield enough hydrogen to replace the vast quantities of fossil fuel used by the transportation sector?

Actually, yes.

According to the folks at fuelcellsworks.com, if the US converted all of its 230 million cars, pickups, vans, and SUVs to hydrogen, they could all be fueled from 310 billion gallons of water each year. While that is a huge quantity of water, it's a drop in the bucket (pun intended) compared to our current consumption. In fact, Americans presently consume over 15 times that amount in their homes — or about 4,800 billion gallons per year — to cook, drink, shower, bathe, flush toilets, and wash clothes. US farmers consume three times *that* amount to irrigate their crops each year. Coal, natural gas, and nuclear power plants use more than 200 times that amount — about 70,000 billion gallons a year. Refineries that extract gasoline from crude oil consume about 300 billion gallons of water.

Clearly, water availability wouldn't limit hydrogen production. What is more, water wouldn't be "consumed" in the process because, as previously noted, hydrogen recombines with oxygen to form water in fuel cells.

What is a Fuel Cell?

Although hydrogen can be burned directly — and cleanly — many experts and government officials believe that the greatest hope for the transportation sector rests on the use of fuel cells. But just what is a fuel cell?

A fuel cell is a rather simple-looking device that combines hydrogen and oxygen to produce electricity. The electricity is then fed to an electric motor that propels a car forward (Figure 9.1).

A fuel cell is an electrochemical device. That is, it produces electricity from fuel, specifically hydrogen and oxygen. The typical fuel cell consists of hundreds of smaller individual fuel cells. They are connected to one another in series, a type of electrical connection that increases voltage. (Each cell puts out about 0.5 to 7.0 volts, depending on current draw). Larger currents are produced by increasing the area of the individual cells.

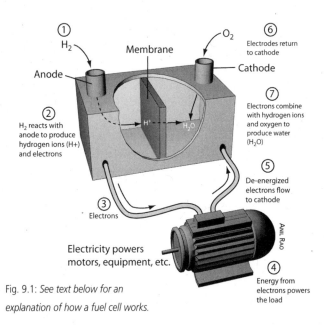

Fig. 9.1: *See text below for an explanation of how a fuel cell works.*

Although there are several types of fuel cells, they all contain an electrolyte sandwiched between two thin electrodes: an anode and cathode. An anode is a negatively charged electrode; a cathode is a positively charged electrode.

In a hydrogen fuel cell, hydrogen is piped into the fuel cell at the anode, which is made of a catalyst, usually a platinum group metal or alloy. (Catalysts are chemicals that speed up chemical reactions without undergoing any chemical change themselves.) Oxygen, which is usually supplied from air, enters at the cathode, which is also made from a catalyst.

At the anode, hydrogen atoms split into negatively charged electrons and positively charged protons. (Remember, a hydrogen atom consists of a single positively charged proton in the nucleus and a negatively charged electron that orbits the nucleus.)

In polymer electrolyte membrane fuel cells (PEM), one common type of fuel cell, positively charged protons move from the anode through the electrolyte to the cathode. When the electrons, protons, and oxygen meet at the cathode side, they combine, producing water and heat.

The movement of the electrons is an important part of the process. As shown in Figure 9.1, electrons are forced to pass from the anode through an external circuit, rather than travelling through the electrolyte. They travel directly from the anode to the cathode by an electrical connection. This flow of electrons creates an electrical current. The energy these electrons "carry" is used to power loads connected to the circuit. (An electrical load is any electrical device that uses electricity.) Because much of the energy the electrons carry is dissipated in loads, the electrons that make it to the cathode are largely de-energized. At the cathode, these electrons recombine with protons and oxygen to produce water. It's all pretty simple — if you have a PhD in electrochemistry, that is.

To summarize, in a hydrogen fuel cell vehicle the anode catalyzes the dissociation (splitting) of hydrogen molecules. This reaction releases the electrons and protons. The electrons flow via an external circuit to the cathode, creating the electrical current. It's this current that's used to power the electric motor that propels the vehicle forward as well as all the smaller loads, such as lights and radio. At the cathode, the electrons, protons, and oxygen react to reform water. Thus direct hydrogen fuel cells produce pure water as their only emission. (For information on other types of fuel cells, see fuelcellsworks.com.)

Hydrogen Cars, Trucks, Vans, and SUVs

You may be surprised to learn that numerous vehicles — including buses, trains, golf carts, bicycles, motorcycles, ships, planes, and even submarines — have been built to run on hydrogen. There's even a hydrogen-powered wheelchair! Hydrogen vehicles can be powered *directly* by hydrogen by burning it in an internal combustion engine similar to those in conventional gasoline-powered vehicles or *indirectly* by hydrogen via fuel cells, as just described.

In fuel cell vehicles powered by hydrogen, hydrogen is fed into a fuel tank at a special filling station (there are only a few in the US, and they're only for experimental vehicles now on the road). Here, the hydrogen is stored under pressure. When the car is turned on, hydrogen flows from the fuel tank to the fuel cell. The fuel cell generates electricity that is fed to the electric motor.

The Pros and Cons of Fuel Cell Vehicles

Fuel cells and fuel cell vehicles offer many advantages over conventional gasoline or diesel-powered vehicles, many of which have been discussed or mentioned earlier in the chapter. Here's a quick summary:

- Fuel cells are powered by the most abundant element on Earth: hydrogen. Hydrogen can be extracted from one of the most abundant chemical compounds: water. Either fresh or salt water can be used. (The electrolysis of seawater is currently a source of chlorine used in industry; the hydrogen produced by this process could also be captured and put to use.)
- The amount of water required is only a small portion of the water already consumed by society.
- Hydrogen could be generated anywhere, so long as there is a supply of electricity and water. (Solar-powered electrolysis units at filling stations could make hydrogen widely available.)
- Hydrogen fuel cells "burn" cleanly and emit only water. (Fuel cells powered by hydrocarbons such as methanol produce water and carbon dioxide.)
- With proper safety measures, hydrogen can be safely stored.
- Pound for pound hydrogen is the most energy-dense fuel available to humankind.

On the downside, hydrogen gas does not appear naturally in great quantities. Large supplies of it would have to be manufactured.

Although pound for pound hydrogen is the most energy-dense fuel available, hydrogen is a gas, so, volumetrically, it is a low-density fuel. Therefore, it is difficult to store enough hydrogen in a tank to generate the same amount of energy available from conventional liquid fuels, such as diesel and gasoline. This poses a problem for drivers who want to travel 300–500 miles on a single tank of fuel. Researchers currently store hydrogen in vehicles in high-pressure tanks.

Another problem is the lack of infrastructure — for example, fueling stations. For hydrogen to be feasible, some argue that we'd need to create a massive delivery system throughout the world. This, in turn, would require a significant investment in time and money. Costs for large-scale deployment would be substantial, although hydrogen could be generated locally at filling stations by small-scale electrolyzers. (Storing significant amounts of hydrogen at local stations, however, could be problematic.)

Another downside of hydrogen fuel cell vehicles is that they can't operate at temperatures below freezing (32°F). Water in the fuel cell solidifies if the cell and its contents are not kept above the freezing point. However, Honda and GM claim they have developed a fuel system that can operate at -20°F.

The most significant downside of hydrogen — and the factor mostly likely to eliminate it from the mix of green fuels — has to do with efficiency. According to several studies, it is three to four times more efficient to run a vehicle on electricity directly than it is to use electricity to generate hydrogen gas from water to feed a fuel cell to make electricity to power a vehicle. As you can see in Figure 9.2, the grid-to-motor efficiency of an electric car is 86%. The grid-to-motor efficiency of a hydrogen fuel cell car is only about 25%.

More than anything else, this fundamental problem is likely to limit hydrogen as a fuel source. Why go through all the steps of making hydrogen when you can power a car directly from electricity at a much higher efficiency? And why bother, when we already have the technology to meet 90% of all Americans' daily transportation needs with 60-mile range electric vehicles? And why spend millions developing hydrogen, when hybrid or plug-in hybrid technology can satisfy practically all other transportation needs?

In summation, while hydrogen-powered vehicles may seem like a wonderful idea, chances are the technology will never gain prominence <u>unless</u> scientists find ways to generate hydrogen with little or no energy input. Those damn laws of physics could banish hydrogen and hydrogen fuel cell vehicles to the scrap heap of exciting but unrealistic pipe dreams.

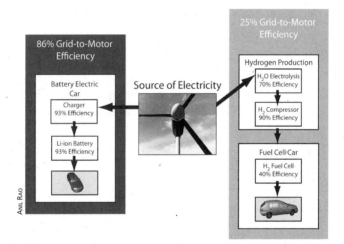

Fig. 9.2: *The most fundamental problem with hydrogen fuel cell vehicles is their inefficiency, as shown here. Bottom line: unless we can find a way to produce hydrogen that requires little energy (good luck!), it makes much more sense to use electricity directly to power a vehicle than to use it to split water to make hydrogen.*

CHAPTER 10

BIOMETHANE

Natural gas-powered vehicles (NGVs) represent yet another green transportation option. Like electric cars, NGVs have been around for a long time. Today, numerous cars, vans, and trucks — especially those in corporate fleets — run on clean-burning natural gas. So do many city buses.

Natural gas is extracted from the Earth's crust where it is found naturally in association with coal and oil deposits or in independent pockets. Natural gas consists mostly of methane gas, though small amounts of other gases, such as carbon dioxide, may also be found in natural gas deposits.

Although it burns cleanly in a car converted to run on natural gas, this fuel is a non-renewable resource; like oil, it has a finite lifespan. So, like oil, methane in natural gas can't be counted on to satisfy our long-term energy needs. It is not a sustainable fuel resource. So why mention natural gas in a book on sustainable transportation?

Methane is produced by numerous natural sources — from flatulent horses and cows to rotting vegetable matter in landfills. It can be made from animal manure.

Humans produce methane as well. As with other animals, gaseous emissions from humans consist of a mixture of methane and carbon dioxide. The gas released by animals and their solid waste is often referred to as *biogas*. Biogas is a mixture of 60% methane and 40% carbon dioxide. Each day, humans produce an amount of biogas equivalent to about 1/200th of a gallon of gasoline. Large quantities can be produced at sewage treatment plants that serve cities with populations in the hundreds of thousands, even millions. In fact, biogas is currently captured at many sewage treatment plants across the world, where it is burned to generate electricity. The electricity is used to run the lights and power pumps and other equipment at the plants. In Norway, biogas from sewage treatment plants is also used to power city buses.

Biogas consisting mostly of methane is also produced in tens of thousands of small methane digesters throughout the less developed world, especially India and China. It is used to power thousands upon thousands of homes, providing light and heat for cooking.

Methane produced from biological sources, referred to as biomethane, *is* a renewable resource. It is clean burning and, like other biofuels, has the potential to help reduce greenhouse gas emissions. For these and other reasons, I've chosen to include biomethane in this book.

Compressed Natural Gas

Most natural gas vehicles in use today operate on compressed natural gas (CNG). Compressing natural gas reduces its volume by 99%. This makes it easier and more economical to store and distribute. CNG is stored in metal tanks at a pressure of about 2,900–3,200 pounds per square inch (psi). (To put this into perspective, air in your car's tires is "stored" at about 30 psi; your back teeth produce 200 psi when biting down on food; an alligator's bite has 2,000–3,000 psi.)

CNG can be burned in traditional gasoline-powered internal combustion engines after modifications are made to the engine, which I'll discuss shortly. Most NGVs, however, are typically bi-fuel vehicles. That is, they burn both gasoline and natural gas.

Although most NGVs on the road today are powered by compressed natural gas, a few of them burn liquefied natural gas (LNG). When this fuel is produced, contaminants like dust, helium, and water must first be removed from natural gas because they can cause problems in engines. The remaining gas is then condensed by being supercooled to approximately -260°F. This reduces its volume considerably, making it practical and less expensive to transport long distances by ship. (LNG is transported in specially designed cryogenic ships or tankers.)

Natural Gas-Burning Cars

NGVs are becoming increasingly popular in Europe and South America. In these regions numerous vehicles now operate on this clean-burning fuel, including cars, vans, pickups, medium-duty delivery trucks, city buses, school buses, and even some trains.

Worldwide, there were an estimated 7 million NGVs operating in 2008. The greatest number of NGVs are in Brazil, Argentina, Italy, Pakistan, Iran, India, and China. South America is the world leader, with nearly half the global market share of NGVs. In Europe, there are about a half a million NGVs. The US is home to about 130,000 NGVs, most of which are city buses.

Several auto manufacturers produce NGVs. The Honda Civic GX is the only commercially available NGV. In Brazil in 2004, GM introduced an engine that could run on compressed natural gas, ethanol, or a gasoline-ethanol blend (E20 to E25). The engine was incorporated into the Chevrolet Astra 2.0 model in 2005 and is used primarily as a taxi. In 2006, the Brazilian subsidiary of the Italian car maker FIAT introduced the Siena Tetra Fuel car. This car runs on natural gas (CNG), 100% ethanol, a gaso-line-ethanol blend (E20 to E25), or pure gasoline.

As noted above, CNG can be burned in traditional gasoline internal combustion engines. It can also be burned in modified diesel engines. But for these vehicles to burn natural gas, they must be equipped with special storage tanks. The tanks are typically made of steel or lighter-weight aluminum, plastic, or a very lightweight composite material.

In addition to the tank, which is located in the trunk, NGVs require pressure regulators. Like the pressure regulators a scuba diver uses, they regulate the flow of gas from the stor-age tank to the engine, reducing pressure so the system operates normally. NGVs also require a gas mixer or gas injector. These deliver gas to the cylinders, much like the injectors in modern gas- or diesel-powered vehicles. The injection system is electron-ically controlled.

NGVs may be filled from either low-pressure ("slow-fill") or high-pressure ("fast-fill") natural gas filling units. (The higher

Natural Gas Combustion

The combustion of biomethane is a chemical reaction. (For those who are scientifically minded, combustion of methane and other organic fuels is considered an oxidation reaction.) During this reaction, oxygen in the air reacts with organic chemicals like methane. The reaction is sparked by a flame. The reaction, in turn, gives off intense heat and light. This, in turn, creates a chain reaction, allowing combustion to self-perpetuate until either the fuel or the oxygen is depleted.

The energy released during this reaction can be used to power a wide assortment of vehicles, but it can also be burned in homes in stoves and ovens, water heaters, furnaces, ☞

the pressure, the faster the tank fills.) Cars can be filled at conventional filling stations equipped with natural gas tanks — though they are rare — or at special facilities that companies maintain to fuel their NGV fleets.

Individuals who own NGVs can also fill their vehicles *at home* if they can get natural gas from a local utility. Fueling a NGV vehicle from a home natural gas line, which is usually located in the garage, is becoming more popular in the US, especially in progressive energy states such as California and New York. Tax credits are available for installing the equipment. A company called FuelMaker has developed a home system in partnership with Honda for the Honda GX.

Sources of Biomethane

Biomethane is produced by the decomposition of organic matter by certain microorganisms in the absence of oxygen. This process is known as *anaerobic digestion*. Anaerobic decomposition produces methane and carbon dioxide, both of which are

and even fireplaces. Methane can also be burned to produce electricity at power plants or generators in industrial facilities. In such cases, natural gas is burned to boil water. High-pressure steam generated during this process is fed through a turbine connected to a generator. The turbine's spinning blades cause the generator to turn, producing electricity.

More commonly, hot, high-pressure exhaust gases from the combustion of methane are fed directly into turbines, which resemble the engines of jets. The high-pressure gases introduced into the turbine cause the blades to spin. The turbine, in turn, powers a generator that produces electricity. ■

combustible. Fuel sources include plant matter, livestock and poultry manure, and human sewage. Even cafeteria and restaurant waste and household garbage can be used to generate methane. But is there enough methane to make it worthwhile?

In the UK, scientists estimate that biogas produced from a variety of sources could replace about 17% of the nation's vehicle fuels. Massive amounts are also available in the US. Many dairy and pig farms in this country, for instance, now pour huge amounts of manure slurry (manure mixed with water) into anaerobic digesters. They crank out methane that farmers use to generate electricity and heat for their operations. Substantial surpluses generated at such facilities are often sold to local utilities to power homes and nearby factories. In the future, massive poultry and livestock operations could become major sources of methane-generated power.

Landfills and sewage treatment plants could also contribute a significant amount of methane. Cafeterias and restaurants around the world produce enormous amounts of organic waste each year. These wastes could be fed into methane digesters to produce biomethane. Even grass clippings could be collected and processed; an acre of grass clippings can produce biomethane containing the energy equivalent of about 300 gallons of gasoline per year.

Although there will not be enough methane to meet all of our needs, no one expects one fuel source to save the day. Biomethane could, however, become a significant source that complements other sustainable fuels like those discussed in this book.

The Pros and Cons of Biomethane

Biomethane is a renewable resource that burns very cleanly. It comes from abundant and diverse sources, and much of our biomethane is currently wasted — it simply escapes into the atmosphere. Once it gets there, it contributes to global warming. In fact, methane is one of the more powerful greenhouse gases produced by human society. Although *burning* methane converts it to carbon dioxide, the latter is a much less powerful greenhouse gas. Therefore, burning biomethane that is being produced anyway could help us reduce global warming that would otherwise occur.

A large point in biomethane's favor is that it can be integrated nicely into the existing infrastructure in many countries. Biomethane produced from waste can be transported in existing natural gas pipelines. Or it can be shipped as LNG, although local consumption of domestic supplies probably makes more sense.

Moreover, methane has many uses besides powering vehicles. It can, for instance, be used in factories as a source of heat and as a source of raw materials. (Natural gas can be used to make fertilizer and hydrogen.) Carbon dioxide stripped from biogas to increase the methane concentration constitutes a valuable chemical in its own right. It can be used to carbonate soft drinks and make dry ice and many other products.

Despite its many advantages, natural gas (all types) faces several problems. Incorporation into the current transportation system will require fuel storage tanks at filling stations throughout the world. Natural gas storage in bi-fuel vehicles also requires an additional storage tank. In cars, these tanks are typically located in the trunks where they take up precious space. Bear in mind, though, that the quantity of natural gas required to power a vehicle is only about one third the volume of hydrogen gas.

Despite these problems, I think the future for natural gas generated from natural sources holds great promise. It could become a major fuel source in the future. Indeed, it probably has to if we are going to continue to prosper on this tiny planet of limited resources. When combined with mass transit and other sensible fuels and vehicles, biomethane could help us create a truly sustainable system of transportation.

Afterword

This book has focused primarily on green personal transportation options: changes in driving habits; clean, sustainably produced, renewable fuels; and environmentally friendly vehicles. My goal has been to help you sort through your options, so you can make wise choices. In this book, I've also offered what I think are the options that make the most sense in both the immediate and the long-term future. By "making sense" I mean they satisfy society's needs for sustainable transportation with fuels and vehicles that are good for people, good for the economy, and good for the environment.

I expect a future with *multiple* green fuels and *multiple* green vehicle options. No single vehicle or fuel will predominate.

In the immediate future, changes in driving habits could go a long way toward helping humanity conserve fuel and reduce pollution — and could save millions of lives as well. I'll admit, though, that I don't think this option holds great promise. Americans are especially resistant to driving more reasonably. Speed seems like a national addiction. Isn't it ironic that we are all in a hurry to get to the couch to chill out!?

Purchasing a more efficient conventional vehicle is another option if you want to take immediate action to green your ride.

For a car, look for mileages in the mid to high 30s or higher — preferably in the 40s and 50s. Anything lower than that just isn't sufficient to help us stretch limited supplies, cut pollution, and create a sustainable future.

Although only briefly mentioned in this book, mass transit — such as high-speed electric trains for cross-country trips and light rail and buses for travel in cities and suburbs — holds the most promise. It is infinitely more efficient than single passenger vehicles, carrying far more people per unit of fuel. Carpools and vanpools can also make a huge contribution to green transportation. Walking and riding bikes are superb options. An added bonus to all these options is reduced congestion, a problem of epic proportions in many parts of the world.

When it comes to commuting, the most promising vehicle is the electric car. Even when powered by electricity from coal-fired power plants, the electric commuter car will help reduce global air pollution, most notably greenhouse gas emissions. Because pollutants are emitted at power plants, electric vehicles reduce our exposure to toxic pollutants that spew from the tailpipes of gasoline- and diesel-powered vehicles. Electric cars operate at a fraction of the cost of conventional vehicles, too. If they can be powered by electricity from wind, solar, hydropower, and geothermal energy, we'd achieve even greater savings — and create a more sustainable system of transportation.

Another commuter car of great promise is the plug-in hybrid electric. It can operate on electricity for up to 60 miles, the range in which 90% of Americans drive in a day. The plug-in hybrid helps reduce emissions, both locally and globally, and if powered by renewably generated electricity, could help us significantly reduce the emissions of the greenhouse gas, carbon dioxide.

Plug-in hybrids offer the benefit of extended range, so a family's commuter car can be used for trips to grandma and grandpa's house for the holidays. If plug-in hybrids were fueled by biodiesel, vegetable oil, biomethane, or ethanol derived from sustainably harvested fuel crops, the environmental benefits would be even greater!

Traditional hybrids, while important in the immediate future, are, in my view, a transitional technology overshadowed by more sustainable vehicles, namely EVs and PHEVs. But I encourage you not to forgo buying a traditional hybrid if you can't find an EV or PHEV that meets your needs and fits your budget.

As for sustainable fuels, I see great promise in electricity generated from renewable sources such as solar and wind. This electricity will serve as a fuel for EVs and PHEVs. I see great promise in cellulosic ethanol and sugar cane ethanol as well. Their net energy efficiency is high, and there's plenty of feedstock for them. Corn ethanol, however, may eventually go the way of the Model T — unless we can improve its production efficiency and tap into cellulose as well.

Biomethane is another highly promising fuel. As you have seen, there are many sources of this gas, and many of them are untapped waste. Methane could be easily piped to homes, where it can be used to power stoves, ovens, water heaters, furnaces, and, ah yes, automobiles and trucks. Filling stations the world over could relatively easily be supplied with biomethane. Both commuters and cross-country drivers could travel using biomethane in a CNG car or truck.

Biodiesel could play a bigger role in the future, as could straight vegetable oil. Only time will tell. I like these fuels and think they offer significant environmental advantages over petro-diesel. If the crops can be grown sustainably or other, more sustainable sources like algae and manure can be tapped to make oil and biodiesel, these fuels could someday be available at many filling stations and become part of the diverse mix of fuels used in a renewably powered society.

These are exciting, and a bit frightening, times, but as you can see there are plenty of solutions that will help us build a green transportation system. Recent changes in the political climate in the US could help "pave" the way to a sustainable transportation system. The Obama Administration's plans for ramping up mass transit, green cars, and green fuels are a step in the right direction, and one that's been needed for a long time. Stay tuned (pun intended).

Resource Guide

Chapter 1 Greening Transportation

Andrews, S. and R. Udall. "The Peal Oil Tank," *Solar Today* 21(3), 31 and 33.

Bartlett, R. G. "Transitioning to a New Paradigm," *Solar Today* 20(2), 27–28.

Ginley, D. and P. Denholm. "Energy Storage: Getting Past the Grid-Lock," *Solar Today* 22(1), 36–39.

Hazard, N. "Driving Change," *Solar Today* 19(4), 28–31.

Heckeroth, S. "Sustainable Transportation: Weighing the Options," *Solar Today* 21(1), 26–27.

Heinbert, R. *The Party's Over: Oil, War and the Fate of Industrial Societies.* New Society Publishers, 2003.

Notari, P. "Addressing the Oil Crisis in the US," *Solar Today* 21(3), 30, 32, and 34.

Chapter 2 Easy Green: Changing Driving Habits, and Other Simple Measures to Green Your Machine

Harvey, G. with S. Prange. "Wise Driving: Outsmart the 7 Worst Gas Guzzlers," *Home Power* 111, 30–33.

Johnson, P. et al., "How to Increase Fuel Mileage on a Car," wikihow.com/increase-fuel-mileage-on-a-car.

Chapter 3 The Hybrid Revolution

Berman, B. "Today's Hybrids," *Home Power* 114, 28–34.

Hammerschlag, R. "Choosing a Green Car," *Solar Today* 19(6), 30–33.

Holan, R. "Hybrids and Diesels Face Off for Fuel Efficient Choices," *Home Power* 117, 48–53.

Masia, S. "Choosing a Low-Carbon Car," *Solar Today* 22(3), 36–41.

Patterson, A. "The 2004 Toyota Prius," *Home Power* 102, 100–104.

Chapter 4 Plug-In Hybrid Electrics

Boschert, S. *Plug-in Hybrids: The Cars That Will Recharge America.* New Society Publishers, 2006.

Boschert, S. "Plug-in Hybrids: Fueling the Future," *Home Power* 121, 56–60.

Duncan, R. "Plug-In Hybrids: Pollution-Free Transport on the Horizon," *Solar Today* 21(3), 46–48.

Hammerschlag, R. "Choosing a Green Car," *Solar Today* 19(6), 30–33.

Prange, S. "The Prius +: A Cleaner Hybrid," *Home Power* 108, 44–48.

Chapter 5 The Electric Car

Brown, M. "Buying a Used Electric Vehicle, Part 1," *Home Power* 89, 110–112.

Brown, M. "Buying a Used Electric Vehicle, Part 2," *Home Power* 90, 120–122.

Brown, M. "Buying a Used Electric Vehicle, Part 3," *Home Power* 91, 98–100.

Brown, M. "Owner's Guide to a Used EV, Part 2," *Home Power* 94, 100–103.

Hammerschlag, R. "Choosing a Green Car," *Solar Today* 19(6), 30–33.

Jensen, M. "PV and EV: My Solar-Electric House and Car," *Home Power* 113, 16–20.

Prange, S. "Charge Your EV, Part 1," *Home Power* 89, 98–100.

Prange, S. "Can an EV Do the Job for Me?," *Home Power* 91, 90–94.

Prange, S. "EV Range: How Much is Enough?" *Home Power* 106, 88–89.

Prange, S. "EV for Sale: Finding and Buying a Used Electric Vehicle," *Home Power* 119, 80–84.

Chapter 6 The Promise and Perils of Ethanol

Cox, J. "Sugar Cane Ethanol's Not-so-sweet Future," August 7, 2007. CNN Money.com.

Freudenberg, R. *Alcohol Fuel: A Guide to Making and Using Ethanol as a Renewable Fuel.* Gabriola Island, BC: New Society Publishers, 2009.

Jacobs, J. "Ethanol from Sugar: What are the Prospects for US. Sugar Co-ops?" rurdev.usda.gov/rbs/pub/sep06/ethanol.htm.

Rohter, L. "With Big Boost from Sugar Cane, Brazil is Satisfying its Fuel Needs," *The New York Times*, nytimes.com/2006/04/10/world/americas/10brazil.html.

Chapter 7 Straight Vegetable Oil

Gregg, F. *SVO: Powering Your Vehicle with Straight Vegetable Oil.* New Society Publishers, 2008.

Journey to Forever. "Straight Vegetable Oil as Diesel Fuel," journey toforever.org/biodiesel_svo.html.

Liess, G. "My Car Runs on Vegetable Oil…Straight Vegetable Oil," *Home Power* 95, 70–74.

Chapter 8 Biodiesel

Alovert, M. "Biodiesel Appleseed," *Home Power* 107, 52–57.

Alovert, M., *Biodiesel Homebrew Guide: Everything You Need to Know to Make Quality Alternative Diesel Fuel from Restaurant Fryer Oil.* You can purchase a copy at localb100.com.

Callaway, D. "Fueling the Biodiesel Debate," *Solar Today* 20(1), 30–33.

Durkee, S. "Getting Off the Petroleum Grid with Biodiesel," *Home Power* 93, 32–39.

Kemp, W. H. *Biodiesel: Basics and Beyond.* Aztext.

Kolod, E. "Going Pro with Biodiesel," *Home Power* 89, 24–29.

Max, D. and R. Engle, "Biofuels: Revolution or Ruse?" *Home Power* 115, 46–49.

Tickel, J. *From the Fryer to the Fuel Tank: The Complete Guide to Using Vegetable Oil as an Alternative Fuel.* Tickel Energy Consultants, 2000.

Chapter 9 Hydrogen and Hydrogen Fuel Cells

Bossel, U. "The Myth of a Hydrogen Future," *Home Power* 114, 82–84.

Engle, R. and D. Crea. "Hydrogen: Solution or Distraction?" *Home Power* 101, 88–93.

FuelCellWorks. "Just the Basics on Hydrogen," fuelcellsworks.com.

FuelCellWorks. "How Fuel Cells Work," fuelcellsworks.com.

FuelCellWorks. "Types of Fuel Cells," fuelcellsworks.com.

Thomas, J., H. Brown and P. Pitchford. "Renewable Hydrogen: A Long-Term Sustainable Solution," *Solar Today* 21(3), 35.

Chapter 10 Biomethane

Blakeslee, T. R. "Biomethane as an Energy Carrier," October 20, 2009, renewableenergyworld.com/rea/news/article/2009/ 10/biomethane-as-an-energy-carrier.

Afterword

Hammerschlag, R. "Choosing a Green Car," *Solar Today* 19(6), 30–33.

Letendre, S. E. "Solar Vehicles at Last?" *Solar Today* 20(3), 26–47.

Prange, S. "Sorting Out the Alternatives: EVs, Hybrids, and Fuel Cell Vehicles," *Home Power* 94, 92–96.

Index

About the Author

D an Chiras is an internationally acclaimed author who has
published over 24 books, including *The Homeowner's Guide
to Renewable Energy* and *Green Home Improvement*. He is a certified wind site assessor and has installed several residential wind
systems. Dan is director of The Evergreen Institute's Center for
Renewable Energy and Green Building (evergreeninstitute. org)
in east-central Missouri, where he teaches workshops on small
wind energy systems, solar electricity, passive solar design and
green building. Dan also has an active consulting business,
Sustainable Systems Design (danchiras.com), and has con-
sulted on numerous projects in North America and Central
America in the past ten years. Dan lives in a passive solar home
powered by wind and solar electricity in Evergreen, Colorado.

If you have enjoyed *Green Transportation Basics*
you might also enjoy other

BOOKS TO BUILD A NEW SOCIETY

Our books provide positive solutions for people who want to
make a difference. We specialize in:

Sustainable Living • Green Building • Peak Oil
Renewable Energy • Environment & Economy
Natural Building & Appropriate Technology
Progressive Leadership • Resistance and Community
Educational and Parenting Resources

New Society Publishers

ENVIRONMENTAL BENEFITS STATEMENT

New Society Publishers has chosen to produce this book on Enviro
100, recycled paper made with **100% post consumer waste**,
processed chlorine free, and old growth free.

For every 5,000 books printed, New Society saves the following
resources:[1]

15	Trees
1,357	Pounds of Solid Waste
1,494	Gallons of Water
1,948	Kilowatt Hours of Electricity
2,468	Pounds of Greenhouse Gases
11	Pounds of HAPs, VOCs, and AOX Combined
4	Cubic Yards of Landfill Space

[1]Environmental benefits are calculated based on research done by the
Environmental Defense Fund and other members of the Paper Task Force who
study the environmental impacts of the paper industry.

For a full list of NSP's titles, please call **1-800-567-6772**
or check out our website at: **www.newsociety.com**

NEW SOCIETY PUBLISHERS